ENERGY SECTOR STANDARD
OF THE PEOPLE'S REPUBLIC OF CHINA

中华人民共和国能源行业标准

Code for Design of High Voltage Electrical Equipment Selection and Arrangement for Hydropower Plants

水力发电厂高压电气设备选择及布置设计规范

NB/T 10345-2019

Replace DL/T 5396-2007

Chief Development Department: China Renewable Energy Engineering Institute
Approval Department: National Energy Administration of the People's Republic of China
Implementation Date: July 1, 2020

China Water & Power Press

Beijing 2024

All rights reserved. No part of this publication may be reproduced, stored in a retrieval system, or transmitted in any form or by any means—electronic, mechanical, photocopying, recording or otherwise, without prior written permission of the publisher.

图书在版编目（CIP）数据

水力发电厂高压电气设备选择及布置设计规范 : NB/T 10345-2019 代替 DL/T 5396-2007 = Code for Design of High Voltage Electrical Equipment Selection and Arrangement for Hydropower Plants (NB/T 10345-2019 Replace DL/T 5396-2007) : 英文 / 国家能源局发布. -- 北京 : 中国水利水电出版社, 2024. 10. ISBN 978-7-5226-2776-2

Ⅰ. TV734

中国国家版本馆CIP数据核字第2024R9V722号

ENERGY SECTOR STANDARD
OF THE PEOPLE'S REPUBLIC OF CHINA
中华人民共和国能源行业标准

Code for Design of High Voltage Electrical Equipment Selection and Arrangement for Hydropower Plants
水力发电厂高压电气设备选择及布置设计规范

NB/T 10345-2019

Replace DL/T 5396-2007

（英文版）

Issued by National Energy Administration of the People's Republic of China
国家能源局　发布
Translation organized by China Renewable Energy Engineering Institute
水电水利规划设计总院　组织翻译
Published by China Water & Power Press
中国水利水电出版社　出版发行
　　Tel: (+ 86 10) 68545888　68545874
　　sales@mwr.gov.cn
　　Account name: China Water & Power Press
　　Address: No.1, Yuyuantan Nanlu, Haidian District, Beijing 100038, China
　　http://www.waterpub.com.cn
中国水利水电出版社微机排版中心　排版
北京中献拓方科技发展有限公司　印刷
184mm×260mm　16开本　5.25印张　166千字
2024年10月第1版　2024年10月第1次印刷
Price（定价）：￥800.00

Introduction

This English version is one of the China's energy sector standard series in English. Its translation was organized by China Renewable Energy Engineering Institute authorized by National Energy Administration of the People's Republic of China in compliance with relevant procedures and stipulations. This English version was issued by National Energy Administration of the People's Republic of China in the Announcement [2023] No.1 dated February 6, 2023.

This version was translated from the Chinese Standard NB/T 10345-2019, *Code for Design of High Voltage Electrical Equipment Selection and Arrangement for Hydropower Plants*, published by China Water & Power Press. The copyright is reserved by National Energy Administration of the People's Republic of China. In the event of any discrepancy in the implementation, the Chinese version shall prevail.

Many thanks go to the staff from relevant standard development organizations and those who have provided generous assistance in the translation and review process.

For further improvement of the English version, any comments and suggestions are welcome and should be addressed to:

China Renewable Energy Engineering Institute
No. 2 Beixiaojie, Liupukang, Xicheng District, Beijing 100120, China
Website: www.creei.cn

Translating organization:

POWERCHINA Chengdu Engineering Corporation Limited

Translating staff:

DU Peilin	MU Kun	QIN Ying	WANG Xinqi
YE Xiuqi	XUE Kaidan	LI Yong	YANG Hong
LIU Bin	LI Yinwei	HUANG Kai	

Review panel members:

GUO Jie	POWERCHINA Beijing Engineering Corporation Limited
LI Zhongjie	POWERCHINA Northwest Engineering Corporation Limited
QIAO Peng	POWERCHINA Northwest Engineering Corporation Limited

ZHANG Ming	Tsinghua University
HOU Yujing	China Institute of Water Resources and Hydropower Research
GAO Yan	POWERCHINA Beijing Engineering Corporation Limited
LI Qian	POWERCHINA Zhongnan Engineering Corporation Limited
CHEN Gang	POWERCHINA Huadong Engineering Corporation Limited

National Energy Administration of the People's Republic of China

翻译出版说明

本译本为国家能源局委托水电水利规划设计总院按照有关程序和规定，统一组织翻译的能源行业标准英文版系列译本之一。2023年2月6日，国家能源局以2023年第1号公告予以公布。

本译本是根据中国水利水电出版社出版的《水力发电厂高压电气设备选择及布置设计规范》NB/T 10345—2019翻译的，著作权归国家能源局所有。在使用过程中，如出现异议，以中文版为准。

本译本在翻译和审核过程中，本标准编制单位及编制组有关成员给予了积极协助。

为不断提高本译本的质量，欢迎使用者提出意见和建议，并反馈给水电水利规划设计总院。

地址：北京市西城区六铺炕北小街2号
邮编：100120
网址：www.creei.cn

本译本翻译单位：中国电建集团成都勘测设计研究院有限公司
本译本翻译人员： 杜沛林　穆　焜　秦　莹　王心琦
　　　　　　　　 叶修齐　薛凯丹　李　勇　杨　红
　　　　　　　　 刘　彬　李寅伟　黄　凯
本译本审核人员：
　郭　洁　中国电建集团北京勘测设计研究院有限公司
　李仲杰　中国电建集团西北勘测设计研究院有限公司
　乔　鹏　中国电建集团西北勘测设计研究院有限公司
　张　明　清华大学
　侯瑜京　中国水利水电科学研究院
　高　燕　中国电建集团北京勘测设计研究院有限公司
　李　倩　中国电建集团中南勘测设计研究院有限公司
　陈　钢　中国电建集团华东勘测设计研究院有限公司

国家能源局

Announcement of National Energy Administration of the People's Republic of China [2019] No. 8

National Energy Administration of the People's Republic of China has approved and issued 152 sector standards including *Code for Operating and Overhauling of Excitation System of Small Hydropower Units* (Attachment 1), and the English version of 39 energy sector standards including *Code for Safe and Civilized Construction of Onshore Wind Power Projects* (Attachment 2).

Attachments: 1. Directory of Sector Standards

2. Directory of English Version of Sector Standards

National Energy Administration of the People's Republic of China

December 30, 2019

Attachment 1:

Directory of Sector Standards

Serial number	Standard No.	Title	Replaced standard No.	Adopted international standard No.	Approval date	Implementation date
...						
22	NB/T 10345-2019	Code for Design of High Voltage Electrical Equipment Selection and Arrangement for Hydropower Plants	DL/T 5396-2007		2019-12-30	2020-07-01
...						

Foreword

According to the requirements of Document GNKJ [2016] No. 238 issued by National Energy Administration of the People's Republic of China, "Notice on Releasing the Development and Revision Plan of the Energy Sector Standards in 2016", and after extensive investigation and research, summarization of practical experience, and wide solicitation of opinions, the drafting group has prepared this code.

The main technical contents of this code include: basic requirements, selection of high voltage electrical equipment, and arrangement of high voltage electrical equipment.

The main technical contents revised are as follows:

—Adding the terms "disassembly-transported transformer" and "take-over current".

—Adding the content of solid insulated tubular bus, disassembly-transported transformer, auxiliary transformer, electronic voltage transformer, electronic current transformer, insulator, bushing, line trap, gas-insulated metal-enclosed switchgear, high voltage fuses.

—Revising the application scope from "3 kV to 500 kV" to "3 kV to 750 kV".

—Revising the elevation correction formula for electrical equipment external insulation, classification of pollution, and the requirements for specific creepage distance.

National Energy Administration of the People's Republic of China is in charge of the administration of this code. China Renewable Energy Engineering Institute has proposed this code and is responsible for its routine management. Energy Sector Standardization Technical Committee on Hydropower Electrical Design is responsible for the explanation of specific technical contents. Comments and suggestions in the implementation of this code should be addressed to:

China Renewable Energy Engineering Institute
No. 2 Beixiaojie, Liupukang, Xicheng District, Beijing 100120, China

Chief development organization:

POWERCHINA Chengdu Engineering Corporation Limited

Participating development organization:

POWERCHINA Northwest Engineering Corporation Limited

Chief drafting staff:

WEN Fengxiang	YANG Hong	HUANG Kai	QIN Ying
ZHAO Huimei	FANG Lei	LI Yong	WANG Yaohui
HOU Yanshuo	SANG Zhiqiang	ZHANG Li	ZHANG Yong
JU Lin	SONG Linli		

Review panel members:

YU Qinggui	FANG Hui	FENG Zhenqiu	KANG Benxian
WANG Jinfu	CHEN Yinqi	SHI Fengxiang	XIA Fujun
YANG Jianjun	SUN Fan	WANG Xiaobing	WANG Yong
SHAO Guangming	WANG Huajun	XIE Xiaohui	DU Gang
LI Shisheng			

Contents

1	**General Provisions**	1
2	**Terms**	2
3	**Basic Requirements**	3
3.1	General Requirements	3
3.2	Electrical Requirements	3
3.3	Mechanical Requirements	5
3.4	Climatic and Environmental Requirements	6
4	**Selection of High Voltage Electrical Equipment**	10
4.1	Conductor	10
4.2	Main Transformer	16
4.3	Auxiliary Transformer	19
4.4	Generator Circuit Breaker	20
4.5	Electric Braking Switch	21
4.6	High Voltage Circuit Breaker	21
4.7	High Voltage Load Switch	22
4.8	High Voltage Disconnector and Earthing Switch	23
4.9	Gas-Insulated Metal-Enclosed Switchgear	23
4.10	AC Metal-Enclosed Switchgear	27
4.11	High Voltage Fuses	28
4.12	Voltage Transformer	28
4.13	Current Transformer	31
4.14	Shunt Reactor	34
4.15	Current-Limiting Reactor	35
4.16	Generator Neutral Equipment	36
4.17	Surge Arrester	39
4.18	Bushing	40
4.19	Insulator	40
4.20	Line Trap	41
5	**Arrangement of High Voltage Electrical Equipment**	42
5.1	Arrangement of Main Electrical Equipment	42
5.2	Clearance Requirements	48
5.3	Access and Fence	60
5.4	Requirements for Buildings and Civil Structures	62
Explanation of Wording in This Code		65
List of Quoted Standards		66

1 General Provisions

1.0.1 This code is formulated with a view to standardizing the selection and arrangement of high voltage electrical equipment for hydropower plants, to achieve the objectives of safety, reliability, technological advancement, and economic rationality.

1.0.2 This code is applicable to the selection and arrangement of high voltage electrical equipment with a nominal voltage of 3 kV to 750 kV for construction, renovation and extension of hydropower plants.

1.0.3 In addition to this code, the selection and arrangement of high voltage electrical equipment for hydropower plants shall comply with other current relevant standards of China.

2 Terms

2.0.1 incoming line

line led from main transformer to high voltage switchgear

2.0.2 outgoing line

line led from high voltage switchgear to high voltage take-off equipment

2.0.3 connecting line

line led from switchgear of a certain voltage level to an interconnecting transformer, or line between switchgear of the same voltage level at different locations

2.0.4 disassembly-transported transformer

power transformer that needs to be disassembled after factory tests and to be reassembled on site due to the transport limitation

2.0.5 take-over current

current at the intersection of the time-current characteristic curves of two kinds of overcurrent protection devices

3 Basic Requirements

3.1 General Requirements

3.1.1 The selection and arrangement of high voltage electrical equipment shall be conducted according to the project-specific conditions and long-term development. Safe, reliable, technologically advanced and economically reasonable products shall be selected. The arrangement scheme shall be formulated according to the requirements of climate, environment, geology, topography, project layout, incoming and outgoing line mode and equipment manufacturing, operation, maintenance, installation and transportation. The selection and arrangement shall follow the principle of land saving and energy conservation, meet the environmental protection requirement, and achieve the objectives of reasonable design, reliable operation, and convenient installation and maintenance.

3.1.2 High voltage switchgear of 72.5 kV or above should be gas-insulated metal-enclosed switchgear (GIS), or may be air-insulated switchgear (AIS) or hybrid gas-insulated switchgear (H-GIS) depending on the project-specific conditions.

3.1.3 The type selection of incoming line, outgoing line and connecting line shall comprehensively consider the main electrical connection, main transformer layout, and type and layout of switchyard.

3.2 Electrical Requirements

3.2.1 The maximum voltage of high voltage electrical equipment shall not be lower than the highest voltage of the system.

3.2.2 The long-term allowable current of the high voltage electrical equipment shall not be less than the maximum continuous working current of the circuit under various possible operation modes. The outdoor high voltage electrical equipment shall consider the effect of sunlight on its current carrying capacity.

3.2.3 The check calculation of short-circuit current shall meet the following requirements:

   1 The short-circuit current used to verify the peak and short-time withstand currents of the high voltage electrical equipment and the breaking current of the circuit-breaker shall be calculated according to the short-circuit current of the power system of the design target year. The main electrical connection and operation mode used in the calculation shall be the normal wiring and operation mode where the

maximum short-circuit current might occur.

2   The short-circuit mode, short-circuit point selection and short-circuit current calculation for the maximum short-circuit current shall comply with the current sector standard NB/T 35043, *Guide for Short-Circuit Current Calculation in Three-Phase AC Systems of Hydropower Projects*.

3   The short-time withstand current of high voltage electrical equipment protected by fuse need not be checked, but the peak withstand current shall be checked. The short-time and peak withstand currents of the circuit of voltage transformer protected by fuse need not be checked.

4   The duration used for calculating the thermal effect of short-circuit current should meet the following requirements:

   1) The duration used for calculating the thermal effect of short-circuit current of conductors other than cables should adopt the sum of the primary protection action time and corresponding circuit breaker break time. When the primary protection has a dead zone, the backup protection action time that can work on dead zone should be adopted.

   2) The duration used for calculating the thermal effect of short-circuit current of cables should adopt the sum of the backup protection action time and corresponding circuit breaker break time. For the feeder cable connected to a motor, the duration used for calculating the thermal effect of short-circuit current may adopt the sum of the primary protection action time and corresponding circuit breaker break time.

   3) The duration used for calculating the thermal effect of short-circuit current of other electrical equipment should adopt the sum of the backup protection action time and corresponding circuit breaker break time.

3.2.4   The frequency of the high voltage electrical equipment shall be determined by the rated frequency of the power system and shall adapt to the system frequency variation range.

3.2.5   The insulation level of the electrical equipment shall comply with the current national standard GB/T 311.1, *Insulation Co-ordination—Part 1: Definitions, Principles and Rules*.

3.2.6   The temperature rise test for high voltage switchgear shall comply with the current sector standard DL/T 593, *Common Specifications for High-Voltage Switchgear and Controlgear Standards*.

3.2.7 Visible corona shall not occur on the electrical equipment and fittings operating under 1.1 times the maximum operating phase voltage at clear nights. The outdoor radio interference voltage of electrical equipment with a voltage of 110 kV or above should not be greater than 500 μV on fine weather days. The test requirements shall comply with the current national standard GB/T 11604, *Testing Procedure of Radio Interference Generated by High Voltage Equipment*.

3.3 Mechanical Requirements

3.3.1 The equipment, supporting structure and its foundation shall be able to withstand the expected mechanical stress. Both normal and abnormal loads shall be considered, and the mechanical strength of the structure shall be determined by the most unfavorable load combination. The load combinations shall meet the following requirements.

1. Under normal load conditions, the static load, tension load, installation load, ice load, wind load, and short-term stress and load during construction and maintenance shall be considered for the equipment, supporting structure and its foundation.

2. Under abnormal load conditions, the simultaneous action of static load, tension load and the maximum accidental load in switching force, short-circuit force, unbalanced tension or seismic load shall be considered for the equipment, supporting structure and its foundation.

3.3.2 The maximum force of the electrical equipment's lead under normal operation and short circuit shall not be greater than the allowable load on the electrical equipment terminal.

3.3.3 The conductor, bushing, insulator and fittings of outdoor switchgear shall be subjected to mechanical calculation based on the meteorological conditions of the power station and different stress states. The minimum safety factors for conductors and insulators shall not be less than those specified in Table 3.3.3.

Table 3.3.3 Minimum safety factors for conductors and insulators

Description	Under long-term load	Under short-term load
Bushing, support insulator and fittings	2.5	1.67
Suspension insulator and fittings	4	2.5

Table 3.3.3 *(continued)*

Description		Under long-term load	Under short-term load
Flexible conductor	Incoming line, outgoing line and connecting line	3.5	2.5
	Others	4	2.5
Rigid conductor		2.0	1.67

NOTES:
1. The safety factor for suspension insulators corresponds to the 1 h electromechanical test load. If corresponding to the failure load, the safety factor shall be 5.3 and 3.3, respectively.
2. The safety factor for rigid conductors corresponds to the failure stress. If corresponding to the yield point stress, the safety factor shall be 1.6 and 1.4, respectively.

3.4 Climatic and Environmental Requirements

3.4.1 The high voltage electrical equipment selected shall meet the service requirements for air temperature, wind speed, humidity, pollution, altitude, earthquake, icing, et.

3.4.2 The selection of ambient temperature for high voltage electrical equipment should be in accordance with Table 3.4.2.

Table 3.4.2 Selection of ambient temperature for high voltage electrical equipment

Description	Type	Ambient temperature	
		Max. temperature	Min. temperature
Bare conductors	Outdoor	Mean maximum temperature of the hottest month	–
	Indoor	Design temperature of ventilation	–
Electrical equipment	Outdoor SF$_6$ insulation equipment	Annual maximum temperature	Extremely low temperature
	Other outdoor equipment	Annual maximum temperature	Annual minimum temperature

Table 3.4.2 *(continued)*

Description	Type	Ambient temperature	
		Max. temperature	Min. temperature
Electrical equipment	Indoor reactor	Design maximum exhaust temperature of ventilation	–
	Other indoor equipment	Design temperature of ventilation	–

NOTES:

1. The annual maximum or minimum temperature shall be the average annual maximum or minimum temperatures measured over years.
2. The mean maximum temperature of the hottest month shall adopt the average value of the monthly average values of the hottest months over years.
3. If there is no data of temperature with ventilation, the ambient temperature for indoor bare conductors and other indoor electrical equipment may take the average maximum temperature in the hottest month plus 5 °C.

3.4.3 When the ambient temperature is lower than the allowable minimum temperature of the high voltage electrical equipment, heating devices or thermal insulation measures shall be taken.

3.4.4 In regions with serious snowing and icing, measures for high voltage electrical equipment shall be taken to prevent accidents due to snow and ice.

3.4.5 For the high voltage electrical equipment with an ambient temperature higher than 40 °C, the test voltage of its external insulation in the dry state shall be multiplied by a temperature correction factor, which shall be calculated by the following formula:

$$K_t = 1 + 0.0033(T - 40) \tag{3.4.5}$$

where

K_t is the temperature correction factor;

T is the ambient air temperature (°C).

3.4.6 The maximum wind speed for selecting the electrical equipment of the outdoor switchgear shall meet the following requirements:

1. For the voltage level of 500 kV or 750 kV, the 10-min mean maximum wind speed with a return period of 50 years measured at 10 m height above the ground should be adopted.

2   For the voltage level of 330 kV or below, the 10-min mean maximum wind speed with a return period of 30 years measured at 10 m height above the ground may be adopted.

3   When the conductor or electrical equipment is at a height more than 10 m above the ground, the maximum wind speed should be corrected according to the installation height of the conductor or electrical equipment, and may be corrected using the relevant formula in the current sector standard DL/T 5158, *Technical Code for Meteorological Survey in Electric Power Engineering*.

4   When the maximum wind speed exceeds 34 m/s, corresponding measures shall be taken for the arrangement of outdoor switchgear.

3.4.7 The relative humidity for the high voltage electrical equipment shall meet the following requirements:

1   When ventilation facilities are set up, the relative humidity shall be selected according to the ventilation design.

2   When ventilation facilities are not provided, the relative humidity of service environment for high voltage electrical equipment shall be the mean relative humidity of the months with the highest humidity over years in the region where the hydropower plant is located.

3   In caverns, underground areas and places with high humidity, the relative humidity of service environment for high voltage electrical equipment shall be the actual relative humidity, or may be taken as 95 % in the absence of relevant data.

3.4.8 The seismic design of high voltage electrical equipment shall comply with the current national standard GB 50260, *Codes for Seismic Design of Electrical Installations*.

3.4.9 For electrical equipment installed at an altitude higher than 1000 m, when the withstand voltage test is conducted on its external insulation, the actual withstand voltage applied to the external insulation shall be multiplied by an elevation correction factor K_a, which shall be calculated by the following formula:

$$K_a = e^{q\left(\frac{H-1000}{8150}\right)} \tag{3.4.9}$$

where

K_a is the elevation correction factor;

H is the elevation of equipment installation site (m);

q is the exponent, which shall comply with the current national standard GB/T 311.1, *Insulation Co-ordination—Part 1: Definitions, Principles and Rules*.

3.4.10 The classification of site pollution and the unified specific creepage distance shall comply with the current national standard GB/T 26218.1, *Selection and Dimensioning of High-Voltage Insulators Intended for Use in Polluted Conditions—Part 1: Definitions, Information and General Principles*.

4 Selection of High Voltage Electrical Equipment

4.1 Conductor

4.1.1 The type of conductor shall be selected according to the requirements of voltage level, loop current, environmental conditions of the power station, site conditions, installation and maintenance, project investment, etc.

4.1.2 The voltage circuit bus of generator may adopt isolated-phase enclosed bus (IPB), common enclosure bus, solid insulated tubular bus, cable, and open bus. The type selection shall meet the following requirements:

1. For the generator voltage circuit bus, when the rated current of the circuit is not larger than 4000 A, common enclosure bus may be selected, and the solid insulated tubular bus may be used when the installation site is narrow. When the rated current of the circuit is larger than 4000 A, connected IPB should be selected. When the rated current of the circuit is not larger than 6300 A and the site is narrow, solid insulated tubular bus may also be used.

2. The main generator voltage circuit bus of bulb tubular turbine units may be cable, common enclosure bus or solid insulated tubular bus.

3. The type of generator voltage circuit bus shall be so selected as to facilitate connection with other generator voltage equipment.

4. When the main generator voltage circuit bus adopts cable, the cable should be of the single-copper-core cross-linked polyethylene (XLPE) type. The voltage between cable core and insulation shield or metal sheath shall be 173 % of the working phase voltage of the circuit. Other technical requirements shall comply with the current national standard GB 50217, *Standard for Design of Cables of Electric Power Engineering*.

5. When the generator voltage circuit adopts open bus, the rectangle bus may be used for 4000 A or below, and the groove-type bus may be used for 4000 A to 8000 A.

6. The generator voltage branch circuit bus should be of the same type as the main circuit bus.

4.1.3 The conductor for high voltage switchgear of 110 kV or above may adopt high voltage cable, gas-insulated metal-enclosed transmission line (GIL), tubular bus, flexible conductor, etc. For incoming lines, outgoing lines, and connecting lines, cable or GIL may be adopted through techno-economic comparison when overhead line is difficult to install. The techno-economic comparison shall consider the conductor's current carrying capacity, civil structures, operation and maintenance, investment, etc.

4.1.4 The selection of IPB shall meet the following requirements:

1. The selection of technical parameters of IPB shall comply with the current national standard GB/T 8349, *Metal-Enclosed Bus*.

2. Temperature measuring devices may be provided at the joints and other locations prone to overheating to monitor the temperature of conductors, joints and enclosures of the IPB.

3. At the connection between the IPB enclosure and the surge arrester cabinet, voltage transformer cabinet or neutral equipment cabinet, an insulating bushing or barrier shall be set to prevent the fault in the cabinet from affecting the bus.

4. Detachable joint shall be applied to the enclosure connection between IPB and equipment other than the generator circuit breaker (GCB) and the phase-change switch. The enclosure of IPB shall be insulated and vibration isolated from the enclosure of equipment.

5. The compensation devices for thermal expansion and contraction or foundation settlement shall be set at the long-straight section, connection section of different foundations and equipment connection of IPB. Its conductor may adopt copper braided, thin aluminum or copper laminated expansion joint, or other equivalent connection modes, and the enclosure may adopt rubber expansion sleeve, aluminum corrugated pipe or other equivalent connection modes.

6. IPB may be equipped with anti-condensation devices at appropriate positions. Self-cooled metal-enclosed bus shall be equipped with the sealing insulation sleeve or other measures at the inner and outer positions through the wall, to prevent the condensation caused by internal and external air convection on the enclosure.

7. The three-phase short-circuit test device shall be provided for the IPB, and shall be arranged to facilitate the generator short-circuit test.

8. Short circuit plates shall be provided on IPB enclosure at the connection to three-phase short-circuit test device and at the connection to the equipment except GCB.

9. The enclosure of IPB and the metal parts of the supporting structure shall be reliably earthed.

10 The enclosure of the continuous IPB may be earthed at multiple points or at one point. The earthing conductor shall meet the requirements of peak withstand current and short-time withstand current.

11 When the short-circuit current passes through the bus, the inductive voltage of the enclosure shall not exceed 24 V.

4.1.5 The selection of the common enclosure bus shall meet the following requirements:

1 The selection of technical parameters of the common enclosure bus shall comply with the current national standard GB/T 8349, *Metal-Enclosed Bus*.

2 Self-cooling shall be adopted for the common enclosure bus.

3 The common enclosure bus with a rated current greater than 2500 A should adopt an aluminum enclosure.

4 The outdoor installed common enclosure bus shall be able to withstand the wind, rain and sunshine without affecting the continuous operation of the bus. No part of the enclosure shall be waterlogged, and the enclosure shall be provided with necessary filters for screening and drainage.

5 The enclosure of the common enclosure bus shall not be less than 4 mm in thickness and shall have enough strength to avoid deformation due to installation error.

6 The common enclosure bus shall be able to compensate for the settlement and displacement of less than 50 mm caused by temperature change and foundation differential settlement of the conductor and enclosure. The expansion joints shall be provided on the straight section of common enclosure bus at an interval of 20 m and at the connection between different foundation structures.

7 The detachable part, cover and outdoor part of inspection hole of the enclosure shall be provided with rubber seal rings. When the enclosure or its supporting structure is displaced by vibration, temperature change or short-circuit electrical force, the ring shall not be displaced or damaged, and the seal ring shall allow for long-term use without replacement or adjustment.

8 The enclosures of common enclosure bus and equipment shall be insulated from each other and adopt detachable connection.

4.1.6 The selection of solid insulated tubular bus shall meet the following requirements:

- 1 The technical parameters of solid insulated tubular bus shall be selected in accordance with the current sector standard DL/T 1658, *Solid Insulated Tubular Bus-Bar of Voltage up to and Including 35 kV.*

- 2 The metal shield of solid insulated tubular bus system shall be insulated in sections, and each section shall be earthed at single point.

- 3 The supports, hangers and fittings of the solid insulated tubular bus shall be made of non-magnetic materials.

- 4 Expansion joints shall be set for each phase of solid insulated tubular bus, and flexible connection shall be provided at the connection with other equipment.

4.1.7 The selection of open bus shall meet the following requirements:

- 1 The open bus shall be selected in accordance with the current sector standard DL/T 5222, *Design Technical Rule for Selecting Conductor and Electrical Equipment.*

- 2 Non-energized metal components such as the support insulator base of the open bus, flange of the bushing, and protective net (cover) shall be earthed reliably.

- 3 In addition to the requirements of electrical and mechanical strength, when the working current of the conductor is greater than 1500 A, the supporting steel members and conductor supporting splint of each phase conductor shall not form a closed magnetic circuit; when the working current is greater than 4000 A, measures shall be taken for the adjacent steel members of the conductor to avoid a closed magnetic circuit, or short circuit rings shall be installed.

4.1.8 The selection of high voltage cable shall meet the following requirements:

- 1 The technical parameter selection and laying requirements for high voltage cables shall comply with the current standards of China GB 50217, *Standard for Design of Cables of Electric Power Engineering;* DL/T 5222, *Design Technical Rule for Selecting Conductor and Electrical Equipment;* and DL/T 5228, *Code for Design of AC 110 kV ~ 500 kV Power Cable Systems for Hydro-Power Station.*

- 2 The cable should be of XLPE type.

- 3 The metal sheath of AC single-core power cable shall be directly earthed at least at one end, and the maximum normal induced potential of any non-earthed end shall meet the following requirements:

1) The maximum normal induced potential of AC single-core power cable shall not be greater than 50 V when no safety measures are taken to effectively prevent personnel from touching the metal sheath.

2) The maximum normal induced potential of AC single-core power cable shall not be greater than 300 V when safety measures are taken to effectively prevent personnel from touching the metal sheath.

4.1.9 The selection of tubular bus shall meet the following requirements:

1. The tubular bus shall be selected by the continuous working current and checked by the short-time withstand current.

2. The tubular bus shall have sufficient mechanical stiffness and strength to avoid breeze vibration and end effects.

3. Tubular bus shall be easy to manufacture and install.

4. For the mechanical calculation of outdoor tubular bus, the conductor load combination cases may be in accordance with Table 4.1.9.

Table 4.1.9 Conductor load combinations

Case	Wind speed	Dead load	Down lead load	Ice load	Short-circuit electromotive force	Seismic load
Normal	Wind speed with ice	√	√	√	–	–
Normal	Maximum wind speed	√	√	–	–	–
Short-circuit	50 % of maximum wind speed and not less than 15 m/s	√	√	–	√	–
Earthquake	25 % of maximum wind speed	√	√	–	–	Earthquake load of corresponding magnitude

NOTE √ denotes the load condition which shall be adopted in the calculation.

5. In the design of outdoor tubular bus, the following formula shall be used to check the breeze vibration speed caused by Karman vortex wind:

$$v_{\mathrm{js}} = f\frac{D}{A} \tag{4.1.9}$$

where

> v_{js} is the calculated wind speed resulting in a breeze resonance at a tubular bus (m/s);
>
> f is the inherent frequency of each order of the conductor (Hz);
>
> D is the outer diameter of conductor (m);
>
> A is the frequency coefficient, which may be taken as 0.214.

6   In the case of no icing and wind, the mid-span deflection of the supported-type single aluminum tubular bus should not exceed 0.5 % of the bus span. The deflection of the split-structure aluminum tubular bus should not exceed 0.4 % of the bus span, and should not be greater than 0.5 to 1.0 times the conductor diameter.

7   The setting of tubular bus earthing point shall ensure that the voltage induced by the adjacent live bus on the bus under maintenance does not exceed 50 V.

8   The expansion joint shall be set reasonably considering the support and fixation mode of the tubular bus.

4.1.10 The selection of flexible conductor shall meet the following requirements:

1   The cross section and structure type of flexible conductor shall be selected according to the environmental conditions, circuit load current, corona, radio interference, etc., and shall be checked by the peak withstand current and short-time withstand current.

2   In coastal areas with high salt mist or places where the surrounding gas is corrosive to the conductor, the anticorrosion type aluminum stranded wire should be selected, and its current carrying capacity may be the value of the same type of wire.

3   When the load current is large, the conductor cross section shall be determined through load current calculation. Conductors of 63 kV or below need not be subjected to the corona voltage check, and conductors of 110 kV or above shall be subjected to the corona voltage check.

4   Single aluminum conductor steel reinforced (ACSR) or composite wire composed of ACSR may be used for flexible wire of 220 kV or below,

hollow expanded wires should be used for 330 kV flexible wire, and double split hollow expanded wire should be used for flexible wire of 500 kV or above.

4.1.11 The selection of GIL shall meet the following requirements:

1. The technical parameters of GIL shall be selected in accordance with the current sector standards DL/T 978, *Specification for Gas-Insulated Metal-Enclosed Transmission Lines*; and DL/T 361, *Technical Guide for Usage of Gas-Insulated Metal-Enclosed Transmission Line*.

2. The connection mode of GIL conductor should be plug-in contact mode, and the connection mode of GIL enclosure should be flange connection.

3. The GIL enclosure should adopt the continuous multi-point earthing mode.

4. The type and length of the GIL standard section shall be determined according to the product characteristics of the manufacturer, transportation conditions of power station, site layout and installation method.

5. The structure of the GIL shall be able to compensate for the settlements and displacement caused by the temperature change and foundation differential settlement of the conductor and enclosure.

4.2 Main Transformer

4.2.1 The main transformer technical parameter selection shall comply with the current national standards GB/T 1094.1, *Power Transformers—Part 1: General*; GB/T 1094.2, *Power Transformers—Part 2: Temperature Rise for Liquid-Immersed Transformers*; GB/T 1094.3, *Power Transformers— Part 3: Insulation Levels, Dielectric Tests and External Clearances in Air*; GB/T 1094.4, *Power Transformers—Part 4: Guide to the Lightning Impulse and Switching Impulse Testing—Power Transformers and Reactors*; GB/T 1094.5, *power transformers—Part 5: Ability to Withstand Short Circuit*; GB/T 1094.7, *Power Transformers—Part 7: Loading Guide for Oil-Immersed Power Transformers*; GB/T 1094.10, *Power Transformers—Part 10: Determination of Sound Levels*; and GB/T 6451, *Specification and Technical Requirements for Oil-Immersed Power Transformers*.

4.2.2 The main transformer should be of oil-immersed type.

4.2.3 The selection of main transformer structure type shall meet the following requirements:

1   Three-phase transformer shall be preferred for main transformer; and three-phase combined power transformer, single-phase transformer set or disassembly-transported power transformer may be selected when transportation conditions are limited.

2   For multiple generators-transformer connection, the main transformer should adopt the step-up double-winding transformer; and the transformer with split winding on the low-voltage side may be used when the short-circuit current needs to be limited.

3   For the interconnecting transformer between two levels of step-up voltage bus, when the neutral points of the two voltage systems are all earthed directly, the step-down auto-transformer should be used. The third winding of delta connection shall be set in the auto-transformer.

4   One standby single-phase transformer may be provided for the single-phase transformer set in any of the following circumstances:

   1) The annual utilization hours of a hydroelectric power plant are 4000 h or above, and there are four or more single-phase transformer sets with the same capacity.

   2) There is only one single-phase transformer set in the whole plant, and the shutdown of the transformer set would result in significant power loss.

   3) There is only one interconnecting single-phase transformer set in the whole plant, there is often large transmission capacity between the two levels of step-up voltage systems, and down maintenance for a long time is not allowed.

4.2.4 The rated capacity of the main transformer shall be selected according to the following requirements:

1   The rated capacity of the main transformer shall match that of the connected hydro-generator, and the rated capacity should be selected from the R10 series in the current national standard GB/T 321, *Preferred Numbers—Series of Preferred Numbers*.

2   The rated capacity of the interconnecting transformer shall be determined according to the requirements of active and reactive power transmission between step-up systems of two different voltages under different operation modes, and the capacity shall not be less than that of the maximum unit connected to the two levels of bus.

3   The rated capacity calculation of the main transformer of pumped

storage power station shall include the capacity of the generator mode and the motor mode of the unit connected to the main transformer. For motor mode, the calculated maximum load of the AC auxiliary system and the load of static frequency converter (SFC) shall also be included.

4.2.5 The selection of rated voltage, tapping and voltage regulation of the transformer winding shall meet the following requirements:

1. The rated voltage, tapping mode and voltage regulation range of the transformer winding shall be determined according to the design of the power station connecting to the power system.

2. When determining the voltage regulation range of the main transformer during the design of the pumped storage power station connecting to the power system, the condenser capacity and lead capacity of the unit and the voltage regulation capability in pumping mode shall be fully considered, and the on-load voltage regulation measures should be avoided as far as possible. When necessary, techno-economic comparison shall be done between increasing the motor generator voltage regulation range of the motor generator and adopting on-load voltage regulation transformer. When the transformer is arranged in underground cavern, increasing the voltage regulating range of the motor generator should be adopted in priority.

4.2.6 The selection of short-circuit impedance shall meet the following requirements:

1. The short-circuit impedance shall be selected according to the design of the power station connecting to the power system and electrical equipment.

2. The short-circuit impedance of the double-winding transformer shall be specified according to the principal tapping; and the short-circuit impedance of each pair of windings of multi-winding transformer shall be specified respectively.

4.2.7 The partial discharge measurement method for transformers shall comply with the current national standard GB/T 1094.3, *Power Transformers—Part 3: Insulation Levels, Dielectric Tests and External Clearances in Air.*

4.2.8 The noise level measurement method for transformers shall comply with the current national standard GB/T 1094.10, *Power Transformers—Part 10: Determination of Sound Levels*; and the value of noise level shall comply with the current sector standard JB/T 10088, *Sound Level for 6 kV ~ 1 000 kV Power Transformers.*

4.2.9 The cooling mode of the main transformer shall be determined according to the environment, capacity, installation position, etc. The outdoor transformer should adopt air-cooling or self-cooling mode, and the indoor large capacity transformer shall adopt water-cooling mode.

4.2.10 The earthing mode and earthing requirements of the transformer core and metal structural parts shall comply with the current national standard GB/T 6451, *Specification and Technical Requirements for Oil-Immersed Power Transformers*. The neutral point of the transformer shall have two down leads connected to different branches of the main earthing grid, and each down lead shall meet the requirements of thermal withstand capability.

4.2.11 For the connection between the transformer and the overhead line, the terminal of the bushing shall comply with the current national standard GB/T 5273, *Dimensional Standardisation of Terminals for High-Voltage Apparatus*.

4.2.12 The connection between the transformer and the cable shall comply with the current sector standard DL/T 5228, *Code for Design of AC 110 kV ~ 500 kV Power Cable Systems for Hydro-Power Station*.

4.2.13 The direct connection between the transformer and GIS or GIL shall comply with the current national standard GB/T 22382, *Direct Connection Between Power Transformers and Gas-Insulated Metal-Enclosed Switchgear for Rated Voltages of 72.5 kV and Above*.

4.2.14 For transformers of 330 kV or above directly connected with GIS, the influence of extra fast transient overvoltage generated by the disconnector operation in GIS on the insulation of the transformer winding shall be considered.

4.2.15 The transformer performance shall meet the requirements of the power system for DC magnetic biasing.

4.3 Auxiliary Transformer

4.3.1 The selection of technical parameters of auxiliary transformer shall comply with the current national standards GB/T 1094.1, *Power Transformers—Part 1: General*; GB/T 1094.2, *Power Transformers—Part 2: Temperature Rise for Liquid-Immersed Transformers*; GB/T 1094.3, *Power Transformers—Part 3: Insulation Levels, Dielectric Tests and External Clearances in Air*; GB/T 1094.4, *Power Transformers—Part 4: Guide to the Lightning Impulse and Switching Impulse Testing—Power Transformers and Reactors*; GB/T 1094.5, *Power Transformers—Part 5: Ability to Withstand Short Circuit*; GB/T 1094.7, *Power Transformers—Part 7: Loading Guide for Oil-Immersed Power Transformers*; GB/T 1094.10, *Power Transformers—Part*

10: *Determination of Sound Levels*; GB/T 1094.11, *Power Transformers—Part 11: Dry-Type Transformers*; and GB/T 1094.12, *Power Transformers—Part 12: Loading Guide for Dry-Type Power Transformers*.

4.3.2 Dry-type transformer with enclosure should be used as indoor auxiliary transformer, and oil-immersed transformer may be used as outdoor auxiliary transformer.

4.3.3 Single-phase dry-type transformer should be used as auxiliary transformer connected with IPB branch. Three-phase transformer may be used as auxiliary transformer with current-limiting reactor and circuit breaker on the high voltage side.

4.3.4 The capacity selection of auxiliary transformer shall comply with the current sector standard NB/T 35044, *Specification for Designing Service Power System for Hydropower Station*.

4.3.5 The short-circuit impedance selection of auxiliary transformer shall consider the selection of the electrical equipment of auxiliary system, the voltage level of the motor starting normally or self-starting in groups, and the influence on the voltage adjustment.

4.3.6 The loss of auxiliary transformer shall comply with the current national standards GB 20052, *Minimum Allowable Values of Energy Efficiency and Energy Efficiency Grades for Three-Phase Distribution Transformers*; and GB 24790, *Minimum Allowable Values of Energy Efficiency and Energy Efficiency Grades for Power Transformers*.

4.4 Generator Circuit Breaker

4.4.1 The technical parameters of GCB shall comply with the current national standard GB/T 14824, *High-Voltage Alternating-Current Generator Circuit-Breaker*.

4.4.2 GCB with a rated short-circuit breaking current above 63 kA should use SF_6 GCB. GCB with a rated short-circuit breaking current of 63 kA or below may use vacuum or SF_6 GCB.

4.4.3 GCB shall be selected according to the short-circuit breaking current on the system side and generator side respectively, and the requirements of transient recovery voltage and DC component shall be considered.

4.4.4 GCB shall have out-of-phase current breaking capability.

4.4.5 The selection of GCB shall consider the altitude correction of current carrying capacity and insulation level.

4.4.6 Both sides of the vacuum GCB should be equipped with resistance and capacity absorption devices.

4.5 Electric Braking Switch

4.5.1 Impulse or frequently started hydro-generator should be equipped with electric braking switch.

4.5.2 The electric braking switch may be a circuit breaker or a disconnector capable of breaking and closing the circuit current.

4.5.3 The rated current of electric braking switch shall meet the requirements of stator circuit braking current and short-time operation of generator under short-circuit drying condition. The short-time withstand current shall be selected reasonably and economically according to the unit braking requirements and the structural characteristics of electric braking switch.

4.5.4 The electric braking switch, which is also used as the short-circuit point in generator short-circuit test, shall meet the requirements of generator short-circuit test.

4.6 High Voltage Circuit Breaker

4.6.1 The technical parameters of high voltage circuit breaker shall comply with the current national standard GB/T 1984, *High-Voltage Alternating-Current Circuit-Breakers*.

4.6.2 Vacuum or SF_6 circuit breakers may be selected for high voltage circuit breakers of 40.5 kV or below. SF_6 circuit breakers should be used for high voltage circuit breakers of above 40.5 kV. Dead tank SF_6 circuit breakers should be used for high-earthquake intensity, heavy pollution, high altitude and alpine areas.

4.6.3 When the DC component of the short-circuit current at the installation site of the circuit breaker does not exceed 20 % of the rated short-circuit breaking current amplitude of the circuit breaker, the DC breaking capacity of the circuit breaker may not be checked. When the DC component of the short-circuit current at the installation site of the circuit breaker exceeds 20 % of the rated short-circuit breaking current amplitude of the circuit breaker, the DC component shall be determined according to the DC component time constant of the project and the current national standard GB/T 1984, *High-Voltage Alternating-Current Circuit-Breakers*.

4.6.4 The rated short-time withstand current duration of the circuit breaker shall meet the following requirements:

   1   When the rated voltage of the circuit breaker is 550 kV to 750 kV, the duration shall be 2 s;

2   When the rated voltage of the circuit breaker is 126 kV to 363 kV, the duration shall be 3 s;

3   When the rated voltage of the circuit breaker is 72.5 kV or below, the duration shall be 4 s.

4.6.5   The selection of rated short-circuit making current of the circuit breaker shall meet the following requirements:

1   When the rated frequency is 50 Hz and the standard value of the time constant is 45 ms, the rated short-circuit making current shall be 2.5 times the effective value of the AC component of the rated short-circuit breaking current.

2   Under special working conditions, the rated short-circuit making current shall be 2.7 times the effective value of AC component of rated short-circuit breaking current, which is independent of the rated frequency of the system.

4.6.6   Whether to install closing resistors for circuit breakers with the rated voltage of 363 kV or above shall be determined according to the overvoltage level.

4.7   High Voltage Load Switch

4.7.1   The technical parameters of high voltage load switch shall comply with current national standards GB/T 3804, *High-Voltage Alternating Current Switches for Rated Voltage Above 3.6 kV and Less than 40.5 kV*; and GB/T 14810, *Alternating Current Switches for Rated Voltages of 72.5 kV and Above*.

4.7.2   High voltage load switch may be used for auxiliary branch circuit or ring main unit without circuit breaker.

4.7.3   High voltage load switch should be of SF_6 or vacuum type.

4.7.4   When the high voltage load switch is combined with the fuse, the load switch shall be able to close the maximum cut-off current of the fuse.

4.7.5   The breaking current of the high voltage load switch shall be greater than the transfer current and the take-over current.

4.7.6   The active load breaking capacity and closed-loop current breaking capacity of the high voltage switch shall not be lower than the rated circuit current.

4.7.7   The high voltage load switch shall have the ability of breaking and closing small inductive and capacitive currents.

4.8 High Voltage Disconnector and Earthing Switch

4.8.1 The technical parameters of high voltage disconnectors and earthing switches shall comply with the current national standard GB/T 1985, *High-Voltage Alternating-Current Disconnectors and Earthing Switches*.

4.8.2 The operating mechanism of disconnectors may be of electric type or manual type according to project-specific conditions.

4.8.3 Disconnectors may be of Class M0, M1 or M2 according to the mechanical life and the operation condition associated with the circuit breaker.

4.8.4 According to the short-circuit making capacity requirement, earthing switches may be of Class E0, E1 or E2. According to the mechanical life and operation condition associated with the circuit breaker, earthing switches may be of Class M0, M1 or M2.

4.8.5 When multiple overhead transmission lines with the rated voltage of 72.5 kV or above are arranged nearby, the line earthing switches shall be able to open and close the induced current and ensure the following operating conditions:

1. If one end of the transmission line is not earthed, the earthing switch installed at the other end shall be able to open and close the capacitive current.

2. If one end of the transmission line is earthed, the earthing switch installed at the other end shall be able to open and close the inductive current.

3. The earthing switch shall be capable of continuously carrying capacitive and inductive currents.

4.8.6 For the 40.5 kV or above disconnector capable of switching the bus-transfer current, the rated bus-transfer current shall be 80 % of the rated current of the disconnector, and shall be determined as agreed with the manufacturer if the rated bus-transfer current exceeds 1600 A.

4.8.7 For the 72.5 kV or above disconnector, which is used to open and close the bus charging current, its opening and closing capacity shall meet the actual requirements of the circuit.

4.9 Gas-Insulated Metal-Enclosed Switchgear

4.9.1 The selection of technical parameters for GIS with a rated voltage of 72.5 kV or above shall comply with the current national standard GB/T 7674, *Gas-Insulated Metal-Enclosed Switchgear for Rated Voltages of 72.5 kV and Above*.

4.9.2 The number of GIS circuit breaker breaks shall be determined according to the equipment manufacturing capacity and voltage level. Single break shall be selected for circuit breakers with a rated voltage of 363 kV or below; single break or double breaks may be selected for circuit breakers with a rated voltage of 550 kV or above. Measures against ferro-resonance shall be taken when double breaks are used.

4.9.3 GIS partitioning shall consider the requirements for phased installation, test, operation, fault, maintenance, etc., and shall meet the following requirements:

1 The overhaul of the equipment in one compartment shall not affect the normal operation of other compartments.

2 Internal faults shall be limited to the fault compartment only.

3 Separate compartments shall be set for circuit breakers, voltage transformers, and surge arresters.

4 Separate compartments shall be set at the connections between GIS and transformer, reactor, high voltage cable, SF_6 or air bushing, or other equipment.

5 The SF_6 gas storage capacity of the compartments shall take into account the recovery time and the capacity of the gas recovery device.

4.9.4 The configuration of expansion joints shall meet the following requirements:

1 The expansion joints shall be so arranged to accommodate the uneven settlement of foundation, civil construction error, equipment manufacturing error, installation error, compensative temperature stress, seismic load, temporary displacement during operation of circuit breaker, and micro vibration of transformer or reactor.

2 The configuration scheme shall be determined according to the specific situation of the project and the structure of GIS. Expansion joints should be set in between long busbars and civil structural joints, and shall be set at the connections between GIS branch buses and transformer/reactor.

4.9.5 The terminals connecting GIS bushing and overhead line shall comply with the current national standard GB/T 5273, *Dimensional Standardisation of Terminals for High-Voltage Apparatus*.

4.9.6 The direct connection between GIS and transformer/reactor shall meet the following requirements:

1 Insulation elements shall be set at the connections between GIS and the enclosure of transformer/reactor for isolation, which shall be able to withstand the maximum induced voltage on the enclosure, and withstand the 2 kV power frequency voltage for 1 min. Zinc-oxide varistors shall be set on both sides of the insulation elements.

2 Detachable break shall be set for the conductive circuit at the connection part between the GIS branch bus and the oil/gas bushing of the transformer/reactor, and shall be accommodated in small compartments.

3 The gap after the detachable part is removed shall be able to withstand various test voltages.

4 During normal operation, the small compartment may be connected with the adjacent compartment by a bypass pipe.

5 When GIS is directly connected to the transformer, non-linear resistors may be set parallel to the insulation between the transformer ascending base and the GIS enclosure. The capacity and characteristics of the non-linear resistor may be determined by the GIS manufacturer.

6 When the GIS is put into operation, if it has not been connected with the transformer/reactor, appropriate sealing measures shall be taken at the connections.

4.9.7 The connections between GIS and cables shall meet the following requirements:

1 The interface design and supply scope of cable terminations and GIS shall comply with the current national standard GB/T 22381, *Cable Connections Between Gas-Insulated Metal-Enclosed Switchgear for Rated Voltages Equal to and Above 72.5 kV and Fluid-Filled and Extruded Insulation Power Cables—Fluid-Filled and Dry Type Cable-Terminations*, and shall facilitate installation, operation, maintenance and testing of cable terminations.

2 Insulation elements shall be set between the GIS enclosure and the metal outer sheath of cable for isolation. The insulation elements shall be able to withstand the maximum induced voltage under various operating conditions, and shall be able to withstand the 2 kV power frequency voltage for 1 min. Zinc oxide voltage varistors shall be set on both sides of the insulation elements.

3 Detachable break shall be set for the conductive circuit at the

connection part between the GIS branch bus and the cable terminations, and shall be accommodated in small compartments.

4　The gap after the detachable part is removed shall be able to withstand various test voltages.

5　During normal operation, the small compartment may be connected with the adjacent compartment by a bypass pipe.

4.9.8　The connection between GIS and GIL shall meet the following requirements:

1　The GIS manufacturer should be responsible for the design of the insulation partition of the connection interface between GIS and GIL, and the GIL manufacturer should be responsible for the design of the conductor connector, enclosure connector and seal parts at the connection between the insulation partition and GIL.

2　The insulation partition between GIS and GIL shall be able to withstand the maximum induced voltage under various operating conditions, and shall be able to withstand the 2 kV power frequency voltage for 1 min.

3　Detachable break shall be set for the conductive circuit at the connection part between the GIS branch bus and GIL, and shall be accommodated in small compartments.

4　The gap after the detachable part is removed shall be able to withstand various test voltages.

5　During normal operation, the small compartment may be connected with the adjacent compartment by bypass pipe, or be provided with the explosion-proof membrane.

4.9.9　The configuration of earthing switches shall meet the following requirements:

1　The Class E0 earthing switch may be used to earth the main circuit of the maintenance part reliably.

2　The Class E1 earthing switch may be used for circuits that cannot be pre-determined whether it is live or not.

3　The earthing terminals of some or all earthing switches shall be insulated from earth potential.

4　The fast earthing switch on the line side shall have suitable capability of switching the electromagnetic induction and electrostatic induction according to the coupling intensity of multiple circuit lines on the same rod or adjacent parallel circuit lines.

4.9.10 For GIS with a rated voltage of 363 kV or above, the influence of very fast transient overvoltage (VFTO) caused by switching operation in GIS on equipment insulation shall be avoided by improving the insulation level of the equipment, optimizing the product structure, reducing the residual charge on the equipment, installing the closing resistor on the circuit breaker, installing the opening resistor on the disconnector, etc. The insulation strength of the main transformer or reactor directly connected with the GIS shall be reviewed according to the calculation and analysis results of VFTO.

4.9.11 The design of GIS enclosure earthing point shall ensure that the induced voltage of equipment enclosure, frame and easily accessible parts shall not be greater than 24 V under normal operation conditions, and the induced voltage under fault conditions shall comply with the current national standard GB/T 50065, *Code for Design of AC Electrical Installations Earthing*.

4.9.12 In severe cold regions, measures shall be taken to prevent insulation gas liquefaction in outdoor GIS.

4.10 AC Metal-Enclosed Switchgear

4.10.1 The technical parameters of AC metal-enclosed switchgear with a rated voltage of 40.5 kV or below shall comply with the current national standards GB/T 3906, *Alternating-Current Metal-Enclosed Switchgear and Controlgear for Rated Voltages Above 3.6 kV and up to and Including 40.5 kV*; and GB/T 11022, *Common Specification for High-Voltage Switchgear and Controlgear Standards*.

4.10.2 Effective blocking and isolation measures shall be taken to prevent fire spread between the cabinets, between the bus rooms, and between functional compartments in the cabinet.

4.10.3 The switch cabinet shall be provided with the following "Five Preventions" measures:

1　To prevent accidental opening or closing of circuit breakers.

2　To prevent closing and opening of disconnectors or isolation plugs with load.

3　To prevent live parts from being earthed by accidental closing of earthing switch.

4　To prevent closing of circuit breaker or disconnectors to earthed parts.

5　To prevent access to energized bay.

4.10.4 Gas-filled AC metal-enclosed switchgear should be selected in the case of limited layout site, and may be used in high altitude areas or poor operation and maintenance conditions.

4.11 High Voltage Fuses

4.11.1 The technical parameters of high voltage fuses shall comply with the current national standards GB/T 15166.1, *High-Voltage Alternating-Current Fuses—Part 1: Terminology*; GB/T 15166.2, *High-Voltage Alternating-Current Fuses—Part 2: Current-Limiting Fuses*; GB/T 15166.3, *High-Voltage Alternating-Current Fuses—Part 3: Expulsion Fuses*; GB/T 15166.4, *High-Voltage Alternating-Current Fuses—Part 4: Fuses for External Protection of Shunt Power Capacitors*; GB/T 15166.5, *High-Voltage Alternating-Current Fuses—Part 5: Specification for High Voltage Fuse-Links for Motor Circuit Applications*; and GB/T 15166.6, *High-Voltage Alternating-Current Fuses—Part 6: Application Guide for the Selection of Fuse-Links of High-Voltage Fuses for Transformer Circuit Applications*.

4.11.2 For a high voltage fuse, the rated current value of the fuse cartridge shall not be less than that of the fuse link.

4.11.3 For a high voltage fuse link, the rated current shall be selected according to the fusing characteristics of the protected object.

4.11.4 A current-limiting high voltage fuse should not be used in the power grid whose working voltage is lower than the rated voltage of the fuse.

4.11.5 Expulsion fuses may be classified into A, B and C according to transient recovery voltage (TRV), and the selection shall comply with the current national standard GB/T 15166.3, *High-Voltage Alternating-Current Fuses—Part 3: Expulsion Fuses*.

4.11.6 For the main circuit of the generator not adopting IPB, the auxiliary branch circuit may adopt a fast current-limiting fuse protection device (FUR).

4.12 Voltage Transformer

4.12.1 The technical parameters of voltage transformers shall comply with the current standards of China GB/T 20840.1, *Instrument Transformers—Part 1: General Requirements*; GB/T 20840.3, *Instrument Transformers—Part 3: Additional Requirements for Inductive Voltage Transformers*; GB/T 20840.5, *Instrument Transformers—Part 5: Additional Requirements for Capacitor Voltage Transformers*; GB/T 20840.7, *Instrument Transformers—Part 7: Electronic Voltage Transformers*; and DL/T 866, *Code for Selection and Calculation of Current Transformer and Voltage Transformer*.

4.12.2 The type selection of voltage transformer shall meet the following requirements:

1. Air-insulated switchgear of 110 kV or above should adopt capacitor voltage transformers.

2. When the line is equipped with carrier communication, the line-side capacitor voltage transformer should be combined with the coupling capacitor.

3. Gas-insulated metal-enclosed switchgear should adopt electromagnetic voltage transformers.

4. 66 kV outdoor switchgear should adopt electromagnetic voltage transformers.

5. Indoor switchgear of 3 kV to 35 kV should adopt solid-insulated electromagnetic voltage transformers, and 35 kV outdoor switchgear may adopt solid-insulated or oil-immersed electromagnetic voltage transformers suitable for outdoor environment.

6. Electronic voltage transformers may be selected according to the project-specific conditions.

4.12.3 The selection of the rated primary voltage of a voltage transformer should meet the following requirements:

1. For the three-phase voltage transformer or the single-phase voltage transformer between three-phase system lines, the standard value of rated primary voltage should be the nominal system voltage.

2. For the single-phase voltage transformer between a three-phase system line and ground, or between the system neutral point and ground, the standard value of primary voltage should be $1/\sqrt{3}$ times the nominal system voltage.

4.12.4 The selection of the rated secondary voltage of a voltage transformer shall meet the following requirements:

1. For the single-phase voltage transformer, the single-phase voltage transformer between three-phase system lines, and the three-phase voltage transformer, the standard value of rated secondary voltage shall be 100 V.

2. For the voltage transformer between the phase and ground in a three-phase system, if the standard value of primary voltage is $1/\sqrt{3}$ times the nominal system voltage, the standard value of rated secondary

voltage shall be $100/\sqrt{3}$ V.

4.12.5 The rated voltage of the residual voltage winding of a voltage transformer shall meet the following requirements:

1. For the voltage transformer of the system with effectively earthed neutral, the standard value of rated voltage of the residual voltage winding shall be 100 V.

2. For the voltage transformer of the system with non-effectively earthed neutral, the standard value of rated voltage of the residual voltage winding shall be 100/3 V.

4.12.6 In the case where the generator adopts 100 % earthing protection device of additional DC stator winding, when a voltage transformer is used to inject DC into the stator winding, the neutral point on the primary side of the voltage transformer connected to the generator voltage shall not be directly earthed, or be earthed through capacitors to isolate DC if earthing is required.

4.12.7 The standard accuracy classes of metering voltage transformers should be 0.1, 0.2, 0.5, 1.0, and 3.0. The standard accuracy classes of protecting voltage transformers should be 3P and 6P. The accuracy classes of residual winding should be 3P and 6P.

4.12.8 The frame for the installation of a voltage transformer should be furnished with two down lead conductors which are connected to different sides of the main earthing grid, and each down lead conductor shall satisfy the thermal stability requirements. Down lead conductors shall facilitate the periodical inspection and testing.

4.12.9 Capacitive voltage transformers shall avoid ferro-resonance and meet the following requirements:

1. In the case of a voltage of $0.8U_{pn}$, $1.0U_{pn}$ or $1.2U_{pn}$ and no load, when the short circuit of the secondary terminals of the transformer is suddenly eliminated after short circuit, the peak of the secondary voltage shall be restored within 0.5 s to the value with a relative difference of no more than 10 % from the normal value before short circuit.

2. In the case where the voltage of the system with effectively earthed neutral is $1.5U_{pn}$ or the voltage of the system with non-effectively earthed neutral is $1.9U_{pn}$ and there is no load, when the short circuit of the secondary terminals of the transformer is suddenly eliminated after short circuit, the ferro-resonance shall last no more than 2 s.

4.12.10 For electromagnetic voltage transformers, the following measures shall be taken to prevent ferro-resonance:

1 The electromagnetic voltage transformer with a high saturation point in excitation characteristic shall be selected.

2 In the same system, the voltage transformer neutral earthing should be minimized. Except for the neutral points of the high voltage winding of the voltage transformer on the power supply side, other voltage transformer neutral points should not be earthed.

3 On a bus of 10 kV or below, a star-connection capacitor bank with neutral earthing may be installed or a section of cable may be used to replace the overhead line, and the ground distributed capacitive reactance X_{co} of the system shall be less than 1 % of the excitation reactance X_m of the single-phase winding under the line voltage action of voltage transformer.

4 The open delta winding of voltage transformer may be equipped with a resistor or other special devices to eliminate ferro-resonance. The resistance shall be calculated by the following formula:

$$R_{13} \leq X_m / K_{13}^2 \qquad (4.12.10)$$

where

R_{13} is the resistance of the open delta winding of voltage transformer (Ω);

X_m is the excitation reactance of the single-phase winding of voltage transformer under line voltage action (Ω);

K_{13} is the transformation ratio of primary winding to open delta winding.

5 The neutral point of the high-voltage winding of voltage transformer may be equipped with a single-phase voltage transformer or a resonance elimination device.

6 Voltage transformers in the system with indirectly earthing neutral shall adopt resonance elimination measures and shall be of the fully insulated type.

4.13 Current Transformer

4.13.1 The technical parameters of a current transformer shall comply with the current standards of China GB/T 20840.1, *Instrument Transformers—Part 1: General Requirements*; GB/T 20840.2, *Instrument Transformers—Part 2: Additional Requirements for Current Transformers*; GB 20840.8, *Instrument*

Transformers—Part 8: Electronic Current Transformers; and DL/T 866, *Code for Selection and Calculation of Current Transformer and Voltage Transformer*.

4.13.2 The type selection of current transformer shall meet the following requirements:

1. Current transformer below 35 kV should adopt solid-insulated structure, and current transformer of 35 kV or above may adopt solid-insulated, gas-insulated or oil-immersed structure.

2. The type selection of current transformers for measurement and protection shall comply with the current sector standard DL/T 866, *Code for Selection and Calculation of Current Transformer and Voltage Transformer*.

3. The electronic current transformer may be selected according to the project-specific conditions. The type selection of electronic current transformer shall comply with the current national standard GB/T 20840.8, *Instrument Transformers—Part 8: Electronic Current Transformer*.

4.13.3 The selection of the rated current of a current transformer shall meet the following requirements:

1. The selection of the rated primary current of a current transformer shall comply with the current sector standard DL/T 866, *Code for Selection and Calculation of Current Transformer and Voltage Transformer*.

2. The standard value of the rated secondary current of a current transformer should meet the following requirements:

    1) 1 A should be selected for new hydropower plants.

    2) 5 A may be selected when the original project adopts 5 A or it is necessary to reduce the secondary open-circuit voltage of the current transformer in some cases.

    3) 1 A and 5 A may be both adopted for the current transformers in a hydropower plant.

3. For neutral current transformers with effective earthing, the rated primary current shall be selected to meet the setting value of relay protection, should be taken as 50 % to 100 % of the rated current of the high voltage side of the transformer, and shall meet the specified error limit.

4. The rated primary current of the zero-sequence current transformer

in the neutral point discharge gap circuit of the transformer should be selected as 100 A.

5　Current transformers for zero-sequence differential protection of auto-transformer shall have the same ratio on each side, and may be selected according to the rated current on the medium voltage side.

6　Current transformers for overload protection and measurement on the common winding of auto-transformer shall be selected according to the allowable load current of the common winding.

7　For the current transformers used for transverse differential protection of generators, the selection of primary current shall meet the following requirements:

1) The current transformer for split-phase transverse differential protection on the neutral point side of branch winding shall be selected according to the total current of parallel branches.

2) The current transformers for zero-sequence current transverse differential protection on the connection lines of each neutral point may be selected according to the maximum allowable unbalanced current of the generator, which is taken as 20 % to 30 % of the rated current of the generator.

4.13.4 The selection of current transformers for protection should meet the following requirements:

1　Class TPY current transformers should be selected for system protection with a voltage of 330 kV or above, differential protection for transformers with the high voltage side being 330 kV or above, and differential protection for generators or generator transformer sets of 300 MW or above.

2　Class PR current transformers should be used for the protection of generator-transformer sets of 200 MW to 300 MW.

3　Class PR or P current transformers may be used for system protection with a voltage of 110 kV to 220 kV, differential protection for transformers with the high voltage side being 110 kV to 220 kV, and protection for generators and generator-transformer sets of 100 MW to 200 MW or below.

4　Class P current transformers should be used for protecting the system with a voltage below 110 kV, transformer differential protection with the high voltage side below 110 kV, and generator-transformer units protection of below 100 MW.

5   When the protection device can overcome the transient saturation impact of current transformer, Class P current transformers may be selected according to the specific requirements for the protection device.

6   The rated short-time thermal current of Class TP transformer should not be less than 10 times the rated primary current, and the rated dynamic current should be 2.5 times the rated short-time thermal current.

7   The short circuit durations of current transformers with different voltage levels should not be less than the values specified in Table 4.13.4.

Table 4.13.4　Short circuit durations of current transformers with different voltage levels

Voltage level (kV)	Short circuit duration (s)
≥ 500	2
110 - 330	3
3 - 66	4

4.13.5　The frame for the installation of a current transformer should be furnished with two down lead conductors which are connected to different sides of the main earthing grid, and each down lead conductor shall meet the thermal stability requirements. Down lead conductors shall facilitate periodical inspection and testing.

4.14　Shunt Reactor

4.14.1　The technical parameters of shunt reactors shall comply with the current national standard GB/T 1094.6, *Power Transformers—Part 6: Reactors*.

4.14.2　Three-phase shunt reactors shall adopt the three-phase five-column type, and shunt reactors of 110 kV or above should adopt the single-phase type.

4.14.3　The small reactance value at the neutral point of the shunt reactor shall be selected according to the situation of the power system to meet the requirements of accelerating the secondary arc extinction or suppressing the resonance overvoltage.

4.14.4　The selection of small reactance at the neutral point of a shunt reactor shall meet the following requirements:

1　The secondary arc current shall not exceed 20 A.

2 The zero-sequence current caused by three-phase unbalance of transmission line may be taken as 0.2 % of the maximum working current of the line.

3 The neutral current caused by three-phase reactance unbalance of the shunt reactor may be taken as 5 % to 8 % of the rated current of the shunt reactor.

4.14.5 The insulation level of the small reactance at the neutral point of a shunt reactor shall be determined through calculation according to the project-specific conditions.

4.14.6 The foundation and the fixing method of a shunt reactor shall consider reducing the vibration of the reactor during operation.

4.14.7 Circuit breakers should not be installed in a shunt reactor circuit, but may be installed in the following cases:

1 Two circuits share one shunt reactor.

2 When the shunt reactor is out of service, the overvoltage level is within the allowable range and the shunt reactor needs to be switched for phase and voltage regulation.

3 There are other special requirements for the electric power system.

4.14.8 Bushing type current transformers should be installed on the high voltage side and neutral side of a shunt reactor.

4.14.9 The surge arrester for the line shunt reactor shall be of type used for transmission lines, and the surge arrester for the bus shunt reactor shall be of type used for power station. The arrester shall be installed as close to the reactor as possible.

4.15 Current-Limiting Reactor

4.15.1 The technical parameters of current-limiting reactors shall comply with the current standards of China GB/T 1094.6, *Power Transformers—Part 6: Reactors*; and DL/T 5222, *Design Technical Rule for Selecting Conductor and Electrical Equipment*.

4.15.2 When the short-circuit current is large, the current-limiting reactor may be used in the branch circuit of auxiliary power, the branch circuit of SFC or the sectional circuit of generator bus.

4.15.3 The current-limiting reactor should be of indoor, single-phase, dry type, and air core type.

4.15.4 The dry type air core reactor should be arranged horizontally.

4.15.5 The rated impedance of current-limiting reactor should be selected in accordance with the following requirements:

1   The current-limiting reactor can limit the short-circuit current to the required value.

2   After current limiting, the short-circuit current of auxiliary power and SFC branch circuit should be limited to 25 kA.

4.16   Generator Neutral Equipment

4.16.1 The earthing mode of generator neutral point shall comply with the current sector standard NB/T 35067, *Overvoltage Protection and Insulation Coordination Design Guide for Hydropower Station*.

4.16.2 The technical parameters of generator neutral earthing transformer shall comply with the current national standard GB/T 1094.11, *Power Transformers—Part 11: Dry Type Transformers*; and the technical parameters of generator neutral point arc suppression coil shall comply with the current national standard GB/T 1094.6, *Power Transformers—Part 6: Reactors*.

4.16.3 The high resistance of generator neutral should be selected according to the temporary overvoltage of the sound phase not exceeding 2.6 times the phase-voltage when the generator is single-phase earthed. The resistance may be calculated by the following formula:

$$R = \frac{10^6}{2\pi f C} \tag{4.16.3}$$

where

R is the connection resistance of generator neutral point (Ω);

f is the generator working frequency (Hz);

C is the total capacitance of three-phase to earth of generator-voltage system (μF).

4.16.4 Dry type single-phase distribution transformer should be selected for earthing transformer of generator neutral point.

4.16.5 The rated voltage of the earthing transformer shall be the generator rated voltage. The capacity is related to its working time and can be calculated by the following formulae:

$$S_e = \frac{U_1 I_c}{k_1} \tag{4.16.5-1}$$

$$I_\mathrm{c} = 2\pi f C \frac{U_1}{\sqrt{3}} \times 10^{-3} \qquad (4.16.5\text{-}2)$$

where

S_e is the rated capacity of earthing transformer (kVA);

U_1 is the rated voltage of generator (kV);

I_c is the capacitance current to the ground of generator and the components connected to the lead-out circuit (A);

k_1 is the overload coefficient, which may be taken as per Table 4.16.5 when the data cannot be obtained from the manufacturer.

Table 4.16.5 Accidental overload coefficient of dry-type transformer

Overload coefficient (overload / rated capacity)	1.4	1.5	1.65	1.7	1.75	1.9	2.2	3.0
Overload duration (min)	60	45	32	18	15	10	5	1

4.16.6 When the 100 % earthing protection of generator stator winding adopts 20 Hz power injection protection, the selection of secondary winding rated voltage shall meet its sensitivity requirements.

4.16.7 The connection resistance on the transformer low voltage side shall be calculated by the following formula:

$$R_2 = Rk^2 - \frac{PU_2^2}{S^2} \qquad (4.16.7)$$

where

R_2 is the connection resistance on the transformer low voltage side (Ω);

U_2 is the voltage on low voltage side of transformer (kV);

k is the ratio of secondary voltage to primary voltage of transformer;

P is the load loss of transformer (kW);

R is the connection resistance of generator neutral point (Ω);

S is the rated capacity of transformer (kVA).

4.16.8 The rated insulation level of arc suppression coil shall be the same as that of generator neutral point.

4.16.9 The tap of arc suppression coil shall meet the requirements of stator earthing protection and tuning, and the number of taps should not be less than 5.

4.16.10 For the generator with unit connection, the neutral arc suppression coil should be under-compensated; for the generator with direct wiring, the neutral arc suppression coil should be overcompensated.

4.16.11 The standard capacity with the capacity close to the calculated value should be selected for the arc suppression coil, and the capacity of the arc suppression coil can be calculated by the following formula:

$$Q = KI_c \frac{U_n}{\sqrt{3}} \tag{4.16.11}$$

where

Q is the compensation capacity (kVA);

K is the coefficient, which may be taken as 1.35 for overcompensation, and may be determined by the out-of-resonance degree for undercompensation;

I_c is the capacitance current to earth of generator and the components connected to the lead-out circuit (A);

U_n is the rated voltage of generator (kV).

4.16.12 For the generator whose neutral is earthed through arc suppression coil, the neutral long-duration voltage displacement shall not exceed 10 % of the rated phase voltage of the generator, and the neutral voltage displacement can be calculated by the following formulae:

$$U_0 = \frac{U_{bd}}{\sqrt{d^2 + v^2}} \tag{4.16.12-1}$$

$$v = \frac{I_c - I_L}{I_c} \tag{4.16.12-2}$$

where

U_0 is the neutral voltage displacement (kV);

U_{bd} is the neutral point asymmetric voltage (kV) of the system or generator circuit before the arc suppression coil is connected, which may be taken as 0.8 % of phase voltage;

d is the damping rate, which may be taken as 3 % for overhead lines of 60 kV to 110 kV, 5 % for 35 kV or below overhead lines, and 2 % to 4 % for cable lines;

- v is the out-of-resonance degree, which should be within ±30 %;
- I_c is the capacitive current of system generator circuit (A);
- I_L is the inductance current of arc suppression coil (A).

4.17 Surge Arrester

4.17.1 The surge arrester should be of AC gapless metal oxide type.

4.17.2 The technical parameters of metal oxide arresters shall comply with the current national standard GB/T 11032, *Metal-Oxide Surge Arresters Without Gaps for A.C. Systems*.

4.17.3 The rated voltage of surge arrester can be selected according to the following formula:

$$U_r \geq kU_t \tag{4.17.3}$$

where

- U_r is the rated voltage of surge arrester (kV);
- k is the coefficient of short-circuit fault clearing time. When the fault clearing time is within 10 s, $k = 1.0$, and when the fault clearing time exceeds 10 s, $k = 1.25$;
- U_t is the temporary overvoltage (kV). When selecting the rated voltage of surge arrester, only the temporary overvoltage caused by phase-to-earth fault, load rejection and long-distance line capacitance effect needs to be considered. The effective value of temporary overvoltage may be selected as per Table 4.17.3.

Table 4.17.3 Effective value of temporary overvoltage

Earthing mode	Non-effective earthing system		Effective earthing system		
				330 - 750	
Nominal system voltage (kV)	3 - 10	35 - 66	110 - 220	Bus	Line
Temporary overvoltage (kV)	$1.1\,U_m$	U_m	$1.3\dfrac{U_m}{\sqrt{3}}$	$1.3\dfrac{U_m}{\sqrt{3}}$	$1.4\dfrac{U_m}{\sqrt{3}}$

NOTE U_m is the highest system voltage.

4.17.4 The root mean square (RMS) value of the continuous operating voltage of the surge arrester shall not be less than 0.8 times the rated voltage of the surge arrester, and shall be in accordance with Table 4.17.4.

Table 4.17.4　RMS value of surge arrester continuous operating voltage

Earthing mode	Effective earthing system	Non-effective earthing system			
System nominal voltage (kV)	110 - 750	3 - 10		35 - 66	
Short-circuit fault clearing time (s)	–	≤ 10	> 10	≤ 10	> 10
Continuous operating voltage (kV)	$\geq \frac{U_m}{\sqrt{3}}$	$\geq \frac{U_m}{\sqrt{3}}$	$\geq 1.1 U_m$	$\geq \frac{U_m}{\sqrt{3}}$	$\geq U_m$

NOTE　U_m is the highest system voltage.

4.17.5 For surge arresters used for generator protection, the rated voltage shall not exceed 1.05 times and 1.25 times the rated voltage of generator respectively when the corresponding fault clearing time is less than 10 s and more than 10 s.

4.18　Bushing

4.18.1 The technical parameters of bushing shall comply with the current national standard GB/T 4109, *Insulated Bushings for Alternating Voltages Above 1000 V*.

4.18.2 The creepage distance of 3 kV to 20 kV wall bushing shall meet the requirements of pollution class at the installation site; otherwise the voltage level of the bushing may be increased by one or two levels.

4.18.3 Porcelain bushing should be used for outdoor bushing in areas with high ultraviolet radiation intensity.

4.19　Insulator

4.19.1 The selection of technical parameters and insulation coordination design of line insulators shall comply with the current national standards GB/T 772, *Technical Specifications of Porcelain Element for High Voltage Insulators*; GB/T 26218.1, *Selection and Dimensioning of High-Voltage Insulators Intended for Use in Polluted Conditions—Part 1: Definitions, Information and General Principles*; and GB 50545, *Code for Design of 110 kV ~ 750 kV Overhead Transmission Line*. The selection of technical parameters of the post insulator shall comply with the current sector standard DL/T 5222, *Design Technical Rule for Selection Conductor and Electrical Equipment.*

4.19.2 Glass insulators should not be used for overhead lines in the plant.

4.19.3 The insulation level of insulator in enclosed bus shall be corrected according to the air temperature in the bus. The test voltage under dry condition shall be the rated withstand voltage multiplied by the temperature correction factor K_t, which shall be calculated by Formula (3.4.5) of this code.

4.19.4 The creepage distance of 3 kV to 20 kV outdoor post insulator shall meet the requirements of pollution class at the installation site; otherwise the voltage level of the post insulator may be increased by one or two levels.

4.19.5 Rod post insulator should be used outdoors, and suspension insulator may be used when it needs to be inverted; multi-edge post insulator with inter and external fittings should be used indoors.

4.20　Line Trap

4.20.1 The technical parameters of line traps shall comply with the current national standard GB/T 7330, *Line Traps for A.C. Power Systems.*

4.20.2 The rated continuous current of line trap shall not be less than the maximum working current of the transmission line with traps in series connection given by the electric power system. The rated short-term current flowing through the main coil within the specified time shall not cause thermal damage or mechanical damage.

5 Arrangement of High Voltage Electrical Equipment

5.1 Arrangement of Main Electrical Equipment

5.1.1 The arrangement of high voltage electrical equipment shall be determined with comprehensive considerations on the wiring mode, equipment type, and general layout of the project.

5.1.2 Generator voltage bus may be arranged in horizontal, vertical or inclined shaft mode, and shall:

1. Meet the requirements for installation, operation, maintenance and overhaul.

2. Meet the requirements for connection of bus with hydro generator, main transformer, excitation transformer, switchgear, current transformer and auxiliary transformer.

3. Meet the requirements of bus support fixing mode and load for civil structures.

5.1.3 The arrangement of the complete set of SF_6 GCB shall:

1. Consider the structural characteristics of the GCB and meet the requirements for the installation, operation, maintenance and overhaul access of the GCB, and for the maintenance space and hoisting of arc extinguishing chamber.

2. Meet the requirements of connection between the circuit breaker and the main circuit bus.

3. Consider the influence of dynamic and static loads on civil structures.

5.1.4 The arrangement of main transformer shall meet the following requirements:

1. The transformer should be arranged close to the generator.

2. The arrangement of transformer yard shall comply with the current standard of China GB 50872, *Code for Fire Protection Design of Hydropower Projects*; and DL 5027, *Typical Extinguishing and Protection Regulation of Electrical Equipment*.

3. The layout of transformer shall provide the necessary access, spaces and handling conditions for transformer unloading, emplacement, repair, test, operation, and maintenance.

4   The transformer radiator should be arranged to the open air.

5   Hoisting facilities should be installed on the ceiling of transformer room for installation and maintenance.

6   The transformer pressure release direction shall avoid the inspection access.

7   The assembly site for the disassembly-transported transformers shall be provided as close as appropriate with a transportation access to the transformer's location.

5.1.5 The GIS arrangement shall consider the space and access for its installation, operation inspection, repair, hoisting, and field test and for handling SF_6 gas recovery device and reserve an erection site. The arrangement shall meet the following requirements:

1   GIS within the same bay shall not be arranged across civil structure joints.

2   The maintenance of components in one bay shall not affect the normal operation of other bays.

3   The access for installation, maintenance and inspection shall be set on both sides of GIS, and shall meet the following requirements:

   1) The main access should be set on the circuit breaker side. Its width shall allow for handling the largest equipment unit of GIS and for the handling of the SF_6 gas recovery unit. The width shall not be less than 2 m, and shall not be less than 2.5 m for GIS of 330 kV or above.

   2) The width of inspection access shall allow for the operation, inspection and gas supply to compartments, and should not be less than 1.0 m. For confined locations, the width shall not be less than 0.8 m.

4   When GIS is arranged indoors, the lifting and handling space shall allow for the installation and maintenance of the largest transportation unit, and the erection site shall be set at one end of the room, the length of which should be 2 times to 3 times the bay width.

5   A control cabinet should be provided for each GIS bay, and should be arranged along the main access of the circuit breaker.

6   The arrangement of GIS shall meet the requirements for the on-site withstand voltage test, and shall consider whether the connection

transition section and the test bushing are needed for the GIS. The test site shall consider the layout, load, transportation and safety clearance of the test equipment.

5.1.6 GIS shall meet the following environmental protection requirements:

1 Mechanical exhaust devices shall be provided for the GIS room. The height of the exhaust outlet to the ground shall be less than 0.3 m, and the exhaust outlet shall be set at a well-ventilated place.

2 The exhaust system of GIS room shall be so designed that SF_6 gas concentration in the GIS room is less than 6.0 g/m^3.

3 A detector for SF_6 gas concentration in the air shall be set in the GIS room. When SF_6 gas concentration in the air exceeds the limit, the detector shall send an alarm signal. Fixed type detector should be selected.

4 The hatch and the cable trench through the wall for the GIS shall be blocked to prevent the leaked SF_6 gas from entering other places with equipment.

5 When the GIS passes through the roof and wall to the outdoor, the seals against rain water shall be set.

5.1.7 GIS room shall meet the following requirements:

1 The indoor GIS room with a floor area of more than 250 m^2 shall have one exit at each end of the room.

2 The underground GIS room shall have moisture-proof measures, and shall be free of groundwater leakage. If necessary, the anti-leakage partition wall and waterproof ceiling shall be set.

3 The allowable deviation for civil structures of GIS room shall meet the following requirements:

1) The allowable horizontal and longitudinal displacement on both sides of concrete foundation joint is ±10 mm, and the allowable vertical displacement is ±5 mm.

2) The allowable deviation of GIS setting surface in the horizontal and vertical directions is ±8 mm.

3) The unevenness of the floor surface is ±10 mm.

4) During GIS operation, the differential settlement of the foundation shall not be greater than 10 mm.

4   Loads on GIS foundation shall mainly include the static load of the equipment, the dynamic load during the operation of the circuit breaker, and the seismic load.

5.1.8 When adopting high voltage cables, the incoming line, outgoing line and connecting line shall meet the following requirements:

1   Cables should not be connected in parallel.

2   Cable should not have intermediate connector.

3   The layout of cable terminals shall meet the following requirements:

   1) The bracket for the cable terminal shall be convenient for the cable passing and hoisting of the cable terminal and its accessories.

   2) When the working current is greater than 1500 A, the steel bracket should not form a closed magnetic circuit around the cable.

   3) The metal sheath earthing junction box should be arranged on the cable terminal bracket. Sheath protectors shall be arranged in the ground junction box or in a position beyond the reach of people. The coaxial cable connecting the metal sheath and earthing junction box shall be as short as possible. The cross section of coaxial cable shall meet the short-time withstand current requirements and the insulation level shall be the same as that of the sheath.

   4) When the indoor cable terminal is used as an applied voltage point for on-site withstand voltage test, sufficient test space shall be reserved according to the test requirements.

4   Cable laying shall meet the following requirements:

   1) The cable route shall be away from oil depot and diesel generator room.

   2) The cable route should be short and easy for cable laying and maintenance.

   3) The cable route should have fewer bends which shall meet the allowable cable bending radius.

   4) The cable route should avoid the place to be excavated.

   5) Tunnels, shafts, inclined shafts or trenches may be set according to the engineering conditions.

   6) Direct burying should not be adopted for cable laying in the plant area.

7) The cable laying shall reserve cable and space for termination.

8) The selection of cable laying mode shall consider the influence on the current carrying capacity and fixed frame of the cable.

5 The cable laying mode may be straight line laying, snake laying or horizontal suspension laying.

6 Cable supporting and fixation shall meet the following requirements:

1) The cable support shall be able to withstand the loads of the cable and its fixings and the loads of longitudinal tension, cable transfer device and human body weight during installation.

2) When the cable support forms a ring, non-magnetic material shall be used for partition.

3) Slip fixed cable lines shall be provided with movable supports.

4) The fixture for the single-core cable shall be non-magnetic material and have the mechanical strength to withstand the action of short-circuit current of the circuit.

5.1.9 The design of overhead incoming line, outgoing line and connecting line shall meet the following requirements:

1 In the design of overhead lines, anchor hooks on hydraulic structures such as dam and powerhouse, or on cliff may be used as wire hanging points.

2 When there is limitation to arrange multiple overhead lines, double circuit lines on the same tower may be used, with overall consideration of earthing design.

3 The influence of abnormal airflow and water mist on the overhead line shall be considered when it crosses the discharge atomization area.

4 The influence on the line shall be considered when overhead line crosses the gate crane, cable crane for construction, navigation facilities and power plant.

5 The overhead line design shall be coordinated with the orientation of the incoming and outgoing lines to avoid line crossing. Overhead line that has to cross other lines due to layout restriction shall meet the following requirements:

1) The overhead line with higher voltage shall be above the line with lower voltage.

2) The overhead line shall take into account the adverse effects of increasing the sag due to the initial wire elongation, wind drift, icing, overload temperature rise, short-circuit overheating, etc., to ensure that the crossing distance meets the electrical requirements in the long-term operation.

6 The minimum distance between overhead lines and the ground or buildings shall comply with the current national standard GB 50545, *Code for Design of 110 kV~750 kV Overhead Transmission Line*.

7 The arrangement of outgoing line gantry should take measures to avoid crossing the expansion joints of structures.

5.1.10 GIL arrangement shall meet the following requirements:

1 The GIL arrangement shall facilitate the connection of GIL to other equipment.

2 A partition shall be set between GIL and GIS.

3 If GIL and GIS need to be tested separately on site, a detachable part should be set between them.

4 The enclosure at the connection between GIL and other equipment shall be insulated.

5 The standard length of GIL shall be determined according to transportation conditions, lifting height and manufacturing technology.

6 The GIL arrangement shall facilitate the installation, inspection, maintenance, overhaul, air replenishment, operation and lifting of GIL and its auxiliary equipment.

7 The GIL arrangement shall meet the GIL field test and site requirements.

8 The GIL arrangement shall take into account the sealing and discharging measures against SF_6 gas leakage and shall be equipped with necessary monitoring devices.

9 For GIL installed in a long vertical or inclined shaft, the influence of elevator, installation and maintenance platform, installation and maintenance of lifting equipment, arrangement of charging and discharging device, ventilation design and cable passage shall be considered.

5.1.11 For the arrangement of AC metal-enclosed switchgear, the structural type of the switchgear and the routing and type of incoming and outgoing lines shall be considered.

5.2 Clearance Requirements

5.2.1 For outdoor switchgear, the allowable minimum safety clearances A'_1, A''_1, A_1, A_2, B_1, B_2, C and D (Figures 5.2.1-1 to 5.2.1-6) should be determined based on the protection level of metal oxide surge arrester. Those for 3 kV to 500 kV outdoor switchgear shall not be less than the values listed in Table 5.2.1-1. Those for 750 kV outdoor switchgear shall not be less than the values listed in Table 5.2.1-2.

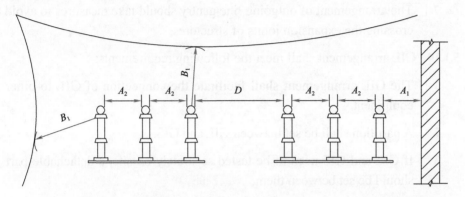

Figure 5.2.1-1   Schematic diagram of safety clearances A_1, A_2, B_1 and D of outdoor switchgear

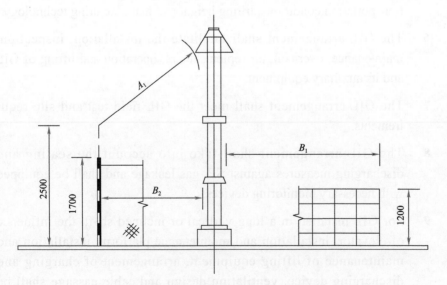

Figure 5.2.1-2   Schematic diagram of safety clearances A_1, B_1 and B_2 of outdoor switchgear (mm)

NB/T 10345-2019

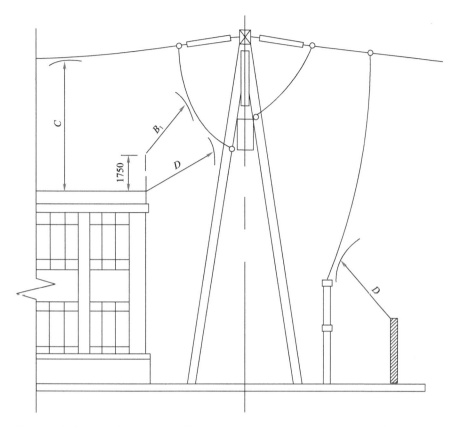

Figure 5.2.1-3　Schematic diagram of safety clearances B_1, C and D of outdoor switchgear (mm)

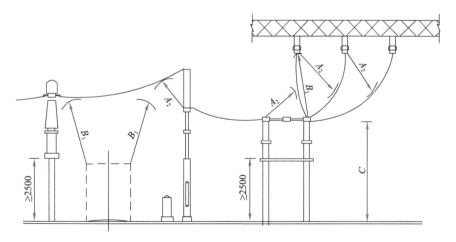

Figure 5.2.1-4　Schematic diagram of safety clearances A_2, B_1 and C of outdoor switchgear (mm)

49

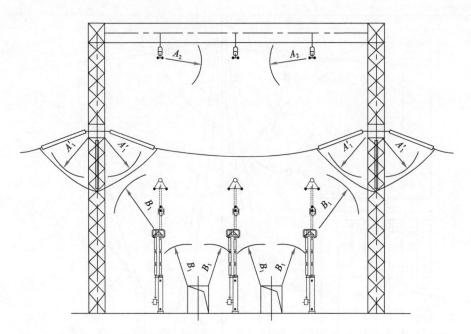

Figure 5.2.1-5 Schematic diagram of safety clearances A'_1, A_2 and B_1 of outdoor switchgear

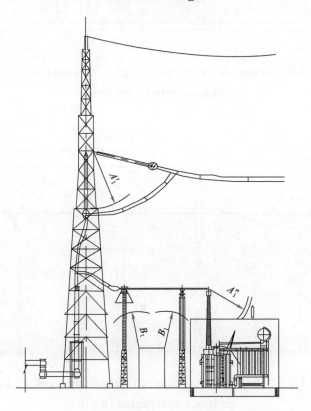

Figure 5.2.1-6 Schematic diagram of safety clearances A'_1, A''_1 and B_1 of outdoor switchgear

Table 5.2.1-1　Minimum safety clearance of 3 kV to 500 kV outdoor switchgear

Safety clearance (mm)	Application scope	System nominal voltage (kV)										Remarks
		3 - 10	15 - 20	35	66	110J	110	220J	330J	500J	
A_1	Between live part and earthed part	200	300	400	650	900	1000	1800	2500	3800	—
	From the net-shaped barrier extension line 2.5 m above the ground to the live parts above the barrier										
A_2	Between the live parts of different phases	200	300	400	650	1000	1100	2000	2800	4300	—
	Between the live leads from either side of the circuit breaker or disconnector										
B_1	Between the outline of passing equipment and the live parts unfenced	950	1050	1150	1400	1650	1750	2550	3250	4550	$B_1=A_1+750$

Table 5.2.1-1 *(continued)*

Safety clearance (mm)	Application scope	System nominal voltage (kV)										Remarks
		3 - 10	15 - 20	35	66	110J	110	220J	330J	500J	
B_1	Between the unfenced live parts that cross each other and are not deenergized simultaneously for maintenance	950	1050	1150	1400	1650	1750	2550	3250	4550	$B_1=A_1+750$
	Between grating fence and insulators or live parts [1]										
B_2	Between net fence and live parts	300	400	500	750	1000	1100	1900	2600	3900	$B_2=A_1+70+30$
	Between unfenced bare conductors and the ground	2700	2800	2900	3100	3400	3500	4300	5000		
C	Between unfenced bare conductors and the top of building and structures									7500[2]	$C=$ $A_1+2300+200$

52

Table 5.2.1-1 (*continued*)

Safety clearance (mm)	Application scope	System nominal voltage (kV)									Remarks
		3 - 10	15 - 20	35	66	110J	110	220J	330J	500J	
D	Between unfenced live parts arranged in parallel that are not deenergized simultaneously for maintenance	2200	2300	2400	2600	2900	3000	3800	4500	5800	$D= A_1+1800+200$
	Between live parts and edges of buildings and structures										

NOTES:
1. 110J, 220J, 330J and 500J refer to systems with effectively earthed neutral.
2. When the altitude exceeds 1000 m, the value of A shall be corrected.
3. For 500 kV, A_1 may be taken as 3500 mm between the twin bundled lines and the earthing part
   ① For switchgear of 220 kV or above, the corresponding value of B_1 may be used for verification according to the actual distribution of the insulator potential. When there is no given distribution potential, it may be calculated according to linear distribution, and the distance between the grating fence and the insulator is allowed to be less than value B_1. The principle may also be used to verify the 500 kV phase-to-phase safety clearance.
   ② The C value of 500 kV switchgear is determined by the field intensity level of electrostatic induction on the ground. The field intensity of the space at 1.5 m from the ground should not exceed 10 kV/m, but may be relaxed to no more than 15 kV/m for particular area.

Table 5.2.1-2 Minimum safety clearance of 750 kV outdoor switchgear

Safety clearance (mm)	Application scope	System nominal voltage (kV) 750J	Remarks
A'_1	Between live conductor and earthing frame	4800	–
A''_1	Between live equipment and earthing frame	5500	–
A_2	Between live conductors	7200	–
B_1	Between live conductor and fence [①]; Between outline of passing equipment and live conductor; Between the unfenced live conductors that cross each other and are not deenergized simultaneously for maintenance	6250	$B_1 = A''_1 + 750$
B_2	Between net fence and live part	5600	$B_2 = A''_1 + 70 + 30$
C	Between live conductor and ground	12000	The value C is determined by the intensity at ground field [②]
D	Horizontal distance between two parallel circuits that are not deenergized simultaneously for maintenance	7500	$D = A''_1 + 1800 + 200$

Table 5.2.1-2 *(continued)*

Safety clearance (mm)	Application scope	System nominal voltage (kV) 750J	Remarks
D	Between live conductor and fence wall top	7500	$D=A_1''+1800+200$
	Between live conductor and building edge		

NOTES:
1. 750J refers to system with effectively earthed neutral.
2. The crossing conductors shall meet the requirements of both A_2 and B_1.
3. Parallel conductors shall meet the requirements of A_2 and D.
4. When the altitude exceeds 1000 m, the value of A shall be corrected.
① For switchgear of 750 kV, the corresponding value of B_1 may be used for verification according to the actual distribution of the insulator potential. When there is no given distribution potential, it may be calculated according to linear distribution, and the distance between the grating fence and the insulator is allowed to be less than value B_1. The principle may also be used to verify the 750 kV phase to phase safety clearance.
② The C value of 750 kV switchgear is determined by the field intensity level of electrostatic induction on the ground. The field intensity of the space at 1.5 m from the ground should not exceed 10 kV/m, but may be no more than 15 kV/m for particular areas.

5.2.2 When flexible conductors are used in the outdoor switchgear, the calculation wind speed and minimum air clearance under different operating conditions from 35 kV to 750 kV shall conform to Table 5.2.2, and the maximum value shall be adopted.

5.2.3 The allowable minimum safety clearances A_1, A_2, B_1, B_2, C, D, and E (Figures 5.2.3-1 to 5.2.3-2) of the indoor switchgear should be determined based on the protection level of the metal oxide surge arrester. The minimum safety clearance of indoor switchgear shall not be less than the value listed in Table 5.2.3.

5.2.4 The safety clearance between live parts of two neighboring switchgears with different system nominal voltages shall be determined according to the higher system nominal voltage.

5.2.5 Lighting, communication or signal lines shall not cross over or pass through above or below the live parts of outdoor switchgear. Exposed lighting and power lines shall not cross over above the live parts of indoor switchgear either.

Table 5.2.2 Calculation wind speed and minimum air clearance under different operating conditions from 35 kV to 750 kV

Condition	Check condition	Calculation wind speed (m/s)	A (mm)	System nominal voltage (kV)							
				35	66	110J	110	220J	330J	500J	750J
Lightning overvoltage	Lightning overvoltage and windage yaw	10[①]	A_1	400	650	900	1000	1800	2400	3200	4300
			A_2	400	650	1000	1100	2000	2600	3600	4800
Switching voltage	Switching overvoltage and windage yaw	50 % of the maximum design wind speed	A_1	400	650	900	1000	1800	2500	3500	4800
			A_2	400	650	1000	1100	2000	2800	4300	6500
Power frequency voltage	1. Maximum operating voltage, short circuit and wind drift (wind speed 10 m/s)	10 or maximum design wind speed	A_1	150	300	300	450	600	1100	1600	2200
	2. Maximum operating voltage and wind drift (maximum design wind speed)		A_2	150	300	500	500	900	1700	2400	3750

NOTE ① For regions with severe climatic conditions where the maximum design wind speed is 34 m/s or above and the high speed wind comes with thunderstorm, the calculation wind speed shall be 15 m/s.

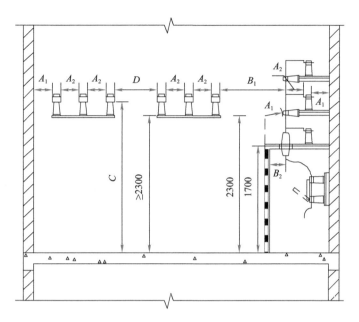

Figure 5.2.3-1 Schematic diagram of A_1, A_2, B_1, B_2, C, D of indoor power switchgear (mm)

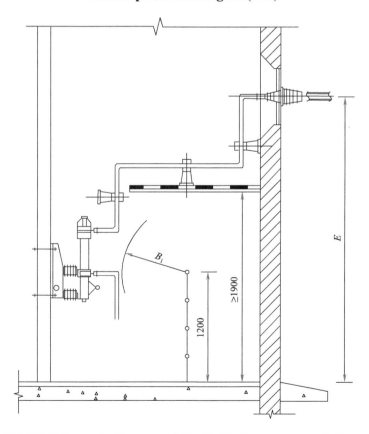

Figure 5.2.3-2 Schematic diagram of B_1, E of indoor power switchgear (mm)

Table 5.2.3 Minimum safety clearance of indoor switchgear

Minimum safety clearance (mm)	Application scope	System nominal voltage (kV)										
		3	6	10	15	20	35	66	110J	220J
A_1	Between live parts and earthed parts	75	100	125	150	180	300	550	850	1800
	Between the net/plate-shaped barrier extension line 2.3 m above the ground to live parts above the barrier									
A_2	Between live parts of the different phases	75	100	125	150	180	300	550	900	2000
	Between live parts of lead at either side of breaks of circuit breakers and disconnectors									
B_1	Between fence-shaped barriers and live parts	825	850	875	900	930	1050	1300	1600	2550
	Cross between the unshielded live parts of different simultaneous outage maintenance									

Table 5.2.3 *(continued)*

Minimum safety clearance (mm)	Application scope	System nominal voltage (kV)								
		3	6	10	15	20	35	66	110J	220J
B_2	Between the network blocks and the live parts	175	200	225	250	280	400	650	950	1900
C	Between the no shading and bare conductor and the floor	2500	2500	2500	2500	2500	2600	2850	3150	4100
D	Parallel between different simultaneous power outages for maintenance of bare conductors	1875	1900	1925	1950	1980	2100	2350	2650	3600
E	The road surface leading to the outhouse from the outgoing casing to the outhouse passage	4000	4000	4000	4000	4000	4000	4500	5000	5500

NOTES:

1 110J, 220J refer to systems with effectively earthed neutral.
2 Between the plate-shaped fence and the live part, the value of B_2 may be (A_1+30) mm.
3 The distance from the switchgear bushing leading out of building to the outdoor ground shall not be less than the value of C for outdoor part listed in Table 5.2.1-1.
4 When the altitude exceeds 1000 m, the value of A shall be corrected.

5.2.6 If the distance from the lowest part of the electrical equipment external insulator of outdoor switchgear to the ground is less than 2.5 m, a fixed fence shall be mounted. If the distance from the lowest part of the electrical equipment external insulator of indoor switchgear to the ground is less than 2.3 m, a fixed fence shall be mounted.

5.3 Access and Fence

5.3.1 The switchgear shall be arranged to allow for easy operation, handling, maintenance, testing and inspection of the equipment, and shall also comply with relevant regulations on safety, fire protection and land saving.

5.3.2 A ring road and patrol path should be set in the main road of the outdoor switchgear of 220 kV or above. Where it is difficult to provide a ring road, turnaround shall be available.

5.3.3 The width of the access to the outdoor switchgear shall meet the following requirements:

1. The main ring access to the outdoor switchgear shall meet the fire protection requirements, and the clear width of the road shall not be less than 4 m.

2. The width of the main access from the gate to the main control building and the main transformer, may be widened to 4.5 m for the 220 kV switchgear, and may be widened to 5.5 m for the 330 kV or above switchgear.

3. The maintenance accesses in the outdoor switchgear and the roads between phases of the outdoor switchgear for 500 kV or above should be 3 m in width.

5.3.4 The road turning radius and slope for the outdoor switchgear should meet the following requirements:

1. The turning radius of the road should not be less than 7 m.

2. The turning radius of the main road should be determined according to the technical performance of the heavy flatbed truck. The turning radius of the main road for the 220 kV switchgear should be 9 m to 12 m, and should be 15 m for the switchgear of 330 kV or above.

3. The longitudinal slope should not be more than 6 % for outdoor switchgear road, and should not be more than 8 % for the roads with a stepped arrangement, in mountainous area or in restricted section.

4. The road should be paved with cement concrete or asphalt.

5.3.5 The patrol access for outdoor switchgear shall meet the following requirements:

1 Patrol accesses shall be set according to the needs of patrol and operation, and patrol route shall be determined considering layout of cable trench on the ground.

2 The width of patrol accesses should be 0.7 m to 1.0 m.

3 The patrol accesses with longitudinal slopes more than 8 % should be provided with anti-slip measures or steps.

5.3.6 The stacked-type outdoor switchgear shall meet the following requirements:

1 The stacked-type outdoor switchgear shall be equipped with upper access and fence.

2 The access width of 220 kV outdoor switchgear may be 3 m to 3.6 m, and the access width of 110 kV outdoor switchgear may be 2 m.

3 A curb of 0.1 m higher than the access shall be set on both sides of the access of outdoor switchgear, and two stairs shall be set; the width of stairs shall not be less than 0.8 m, the slope shall not be more than 45°, and the surface shall be provided with anti-slip measures.

4 An outdoor overpass should be set between two adjacent stacked-type outdoor switchgear, or between the upper access of the stacked-type outdoor switchgear and the control building.

5 An overpass should be provided between the indoor switchgear building and the control building if they are close to each other.

5.3.7 For indoor high voltage switchgear of 35 kV or below arranged in rows, the minimum clear width of access shall meet the requirements in Table 5.3.7.

5.3.8 When the oil-immersed transformer is arranged indoors, for the transformer with a capacity of 1250 kVA or above, the minimum clearance from the outline of oil pond (pit) to the rear or side wall of the transformer room shall not be less than 800 mm, and that to the door shall not be less than 1000 mm. For the transformer with a capacity of 1000 kVA or below, the minimum clearance from the outline of the oil pond (pit) to the rear or side wall of the transformer chamber shall not be less than 600 mm, and that to the door shall not be less than 800 mm.

5.3.9 For indoor dry-type transformers without casing, the clearance between its outline and the surrounding walls shall not be less than 600 mm, and the clearance between dry-type transformers shall not be less than 1000 mm.

Table 5.3.7 Minimum clear width of access (mm)

Arrangement	Access for maintenance	Access for operation	
		Fixed type	Removable type
Single-row	800	1500	Length of a single handcart +1200
Double-row	1000	2000	Length of double handcarts +900

NOTES:
1. The minimum clear width of accesses may be reduced by 200 mm in the case of a protruding portion of a wall or column.
2. For the removable switchgear cabinets not subjected to local maintenance, the width of the accesses may be reduced properly.
3. Where a fixed type switchgear cabinet is arranged close to wall, the distance between the rear side thereof and the wall should be 50 mm.
4. Where a 35 kV handcart switchgear cabinet is used, the access width behind it should not be less than 1000 mm.
5. For the complete set of fully insulated and sealed switchgear, the access width may be reduced according to the installation and operation instructions of the manufacturer.

5.3.10 For the independent switchyard of the air-insulated high voltage switchgear, the fence wall with height no less than 2200 mm shall be set around the site; a fence shall be installed around the outdoor switchgear in the power plant and its height shall not be less than 1500 mm.

5.3.11 The fence for switchgear shall meet the following requirements:

1. The height of grating fence of the electrical equipment shall not be less than 1200 mm, and the clearance from the lowest cross bar to the ground shall not be more than 200 mm.
2. The height of net fence of electrical equipment shall not be less than 1700 mm, and its meshes shall not be larger than 40 mm × 40 mm.
3. The lock shall be provided for the fence gate.

5.3.12 The outdoor bus bridge shall be provided with protective measures to prevent foreign matters from falling on the bus.

5.4 Requirements for Buildings and Civil Structures

5.4.1 The switchgear room shall meet the following requirements:

1. The exits of the switchgear room shall meet the following requirements:

1) The switchgear room with a length of over 7 m shall have two exits, which should be arranged at both ends of the switchgear room.

2) Except GIS room, the switchgear room with a length of over 60 m should have an extra exit in the middle of the room. When the area of GIS room exceeds 250 m^2, an exit shall be set at each end of the room.

3) The switchgear room arranged upstairs shall have at least one emergency exit leading to the corridor on that floor or to the outside.

2 The partition with a door opening should be provided at locations where bus of indoor air-insulated switchgear is sectionalized.

3 Where the door to an oil-filled electrical equipment room opens towards the interior of buildings beyond the scope of switchgear, the door shall be a fire-proof door.

4 The door to the switchgear room shall be a fire-proof door that opens to the direction of evacuation.

5 The switchgear rooms may be provided with openable windows, and measures shall be taken to prevent rain, snow, small animals, wind and dust from entering.

6 The fire endurance rating and decoration of the switchgear room shall comply with the national standard GB 50872, *Code for Fire Protection Design of Hydropower Projects*.

7 For multi-storey switchgear rooms, anti-leakage measures shall be taken for floors.

8 The indoor access in the switchgear rooms shall not be obstructed, doorsill shall not be set and pipes irrelevant to the switchgear shall not pass through.

9 For the switchgear room of 20 kV or below, the door size should allow for the transportation of the maximum size equipment, its clear height should be the equipment height plus 0.5 m and its clear width should be the equipment width plus 0.3 m. The clear width of the evacuation access door shall not be less than 0.9 m and the clear height should not be less than 2.0 m.

10 The floor of the switchgear room should be 100 mm to 300 mm higher than the ground.

5.4.2 The load on the outdoor switchgear framework shall be determined according to the following requirements:

1. Meteorological conditions used for calculation shall be determined based on local meteorological data.

2. The framework should be designed according to the stress conditions, including the possible unfavorable conditions in the future. The terminal and intermediate framework should be designed separately. The framework design need not consider the case of transmission line breaking.

3. The framework design shall consider the load combinations for normal operation, installation, maintenance and earthquake, and should meet the following requirements:

    1) For normal operation conditions, the framework design may adopt the load corresponding to the 50-year maximum wind speed, lowest air temperature or the most severe icing, whichever the most unfavorable.

    2) For installation conditions, the erection of conductors and earth wires shall consider a load of 2 kN produced by the man and tools acting on the cross beam, the corresponding wind load, the tension and self-weight of conductors and earth wires, etc.

    3) For maintenance conditions, the framework design may consider the wire tension, corresponding wind load, self-weight, etc. in the two cases, i.e., manual maintenance on deenergized three phases at the same time, and manual maintenance on energized single phase at the mid-span. In the absense of a down lead within a span, manual maintenance at the mid-span need not be considered.

    4) For earthquake conditions, the framework design shall consider the horizontal earthquake action, the corresponding wind or icing load, self-weight and tension of conductors and earth wires, etc. The adjustment coefficients for structural resistance, uplift resistance, overturning resistance, and bearing capacity shall comply with the current national standard GB 50191, *Code for Seismic Design of Special Structures*.

Explanation of Wording in This Code

1 Words used for different degrees of strictness are explained as follows in order to mark the differences in executing the requirements in this code.

　1) Words denoting a very strict or mandatory requirement:

　　"Must" is used for affirmation, "must not" for negation.

　2) Words denoting a strict requirement under normal conditions:

　　"Shall" is used for affirmation, "shall not" for negation.

　3) Words denoting a permission of a slight choice or an indication of the most suitable choice when conditions permit:

　　"Should" is used for affirmation, "should not" for negation.

　4) "May" is used to express the option available, sometimes with the conditional permit.

2 "Shall meet the requirements of…" or "shall comply with…" is used in this code to indicate that it is necessary to comply with the requirements stipulated in other relative standards and codes.

List of Quoted Standards

GB/T 50064,	*Code for Design of Overvoltage Protection and Insulation Coordination for AC Electrical Installations*
GB/T 50065,	*Code for Design of AC Electrical Installations Earthing*
GB 50191,	*Code for Seismic Design of Special Structures*
GB 50217,	*Standard for Design of Cables of Electric Power Engineering*
GB 50260,	*Codes for Seismic Design of Electrical Installations*
GB 50545,	*Code for Design of 110 kV ~ 750 kV Overhead Transmission Line*
GB 50872,	*Code for Fire Protection Design of Hydropower Projects*
GB/T 311.1,	*Insulation Co-ordination—Part 1: Definitions, Principles and Rules*
GB/T 321,	*Preferred Numbers—Series of Preferred Numbers*
GB/T 772,	*Technical Specifications of Porcelain Element for High Voltage Insulators*
GB/T 1094.1,	*Power Transformers—Part 1: General*
GB/T 1094.2,	*Power Transformers—Part 2: Temperature Rise for Liquid-Immersed Transformers*
GB/T 1094.3,	*Power Transformers—Part 3: Insulation Levels, Dielectric Tests and External Clearances in Air*
GB/T 1094.4,	*Power Transformers—Part 4: Guide to the Lightning Impulse and Switching Impulse Testing—Power Transformers and Reactors*
GB/T 1094.5,	*Power Transformer—Part 5: Ability to Withstand Short Circuit*
GB/T 1094.6,	*Power Transformers—Part 6: Reactors*
GB/T 1094.7,	*Power Transformers—Part 7: Loading Guide for Oil-Immersed Power Transformers*
GB/T 1094.10,	*Power Transformers—Part 10: Determination of Sound Levels*
GB/T 1094.11,	*Power Transformers—Part 11: Dry-Type Transformers*

GB/T 1094.12,	*Power Transformers—Part 12: Loading Guide for Dry-Type Power Transformers*
GB/T 1984,	*High-Voltage Alternating-Current Circuit-Breakers*
GB/T 1985,	*High-Voltage Alternating-Current Disconnectors and Earthing Switches*
GB/T 3804,	*High-Voltage Alternating Current Switches for Rated Voltage Above 3.6 kV and Less than 40.5 kV*
GB/T 3906,	*Alternating-Current Metal-Enclosed Switchgear and Controlgear for Rated Voltages Above 3.6 kV and up to and Including 40.5 kV*
GB/T 4109,	*Insulated Bushings for Alternating Voltages Above 1000 V*
GB/T 5273,	*Dimensional Standardisation of Terminals for High-Voltage Apparatus*
GB/T 6451,	*Specification and Technical Requirements for Oil-Immersed Power Transformers*
GB/T 7330,	*Line Traps for A.C. Power Systems*
GB/T 7674,	*Gas-Insulated Metal-Enclosed Switchgear for Rated Voltages of 72.5 kV and Above*
GB/T 8349,	*Metal-Enclosed Bus*
GB/T 11022,	*Common Specification for High-Voltage Switchgear and Controlgear Standards*
GB/T 11032,	*Metal-Oxide Surge Arresters without Gaps for A.C. Systems*
GB/T 11604,	*Testing Procedure of Radio Interference Generated by High Voltage Equipment*
GB/T 14810,	*Alternating Current Switches for Rated Voltages of 72.5 kV and Above*
GB/T 14824,	*High-Voltage Alternating-Current Generator Circuit-Breaker*
GB/T 15166.1,	*High-Voltage Alternating-Current Fuses—Part 1: Terminology*
GB/T 15166.2,	*High-Voltage Alternating-Current Fuses—Part 2: Current-Limiting Fuses*
GB/T 15166.3,	*High-Voltage Alternating-Current Fuses—Part 3:*

	Expulsion Fuses
GB/T 15166.4,	*High-Voltage Alternating-Current Fuses—Part 4: Fuses for External Protection of Shunt Power Capacitors*
GB/T 15166.5,	*High-Voltage Alternating-Current Fuses—Part 5: Specification for High-Voltage Fuse-Links for Motor Circuit Applications*
GB/T 15166.6,	*High-Voltage Alternating-Current Fuses—Part 6: Application Guide for the Selection of Fuse-Links of High-Voltage Fuses for Transformer Circuit Applications*
GB 20052,	*Minimum Allowable Values of Energy Efficiency and Energy Efficiency Grades for Three-Phase Distribution Transformers*
GB/T 20840.1,	*Instrument Transformers—Part 1: General Requirements*
GB/T 20840.2,	*Instrument Transformers—Part 2: Additional Requirements for Current Transformers*
GB/T 20840.3,	*Instrument Transformers—Part 3: Additional Requirements for Inductive Voltage Transformers*
GB/T 20840.5,	*Instrument Transformers—Part 5: Additional Requirements for Capacitor Voltage Transformers*
GB/T 20840.7,	*Instrument Transformers—Part 7: Electronic Voltage Transformers*
GB/T 20840.8,	*Instrument Transformers—Part 8: Electronic Current Transformers*
GB/T 22381,	*Cable Connections Between Gas-Insulated Metal-Enclosed Switchgear for Rated Voltages Equal to and Above 72.5 kV and Fluid-Filled and Extruded Insulation Power Cables—Fluid-Filled and Dry Type Cable-Terminations*
GB/T 22382,	*Direct Connection Between Power Transformers and Gas-Insulated Metal-Enclosed Switchgear for Rated Voltages of 72.5 kV and Above*
GB 24790,	*Minimum Allowable Values of Energy Efficiency and Energy Efficiency Grades for Power Transformers*
GB/T 26218.1,	*Selection and Dimensioning of High-Voltage Insulators Intended for Use in Polluted Conditions—Part 1:*

Definitions, Information and General Principles

NB/T 35043,	*Guide for Short-Circuit Current Calculation in Three-Phase AC Systems of Hydropower Projects*
NB/T 35044,	*Specification for Designing Service Power System for Hydropower Station*
NB/T 35067,	*Overvoltage Protection and Insulation Coordination Design Guide for Hydropower Station*
DL/T 361,	*Technical Guide for Usage of Gas-Insulated Metal-Enclosed Transmission Line*
DL/T 593,	*Common Specifications for High-Voltage Switchgear and Controlgear Standards*
DL/T 866,	*Code for Selection and Calculation of Current Transformer and Voltage Transformer*
DL/T 978,	*Specification for Gas-Insulated Metal-Enclosed Transmission Lines*
DL/T 1658,	*Solid Insulated Tubular Bus-Bar of Voltage Up to and Including 35 kV*
DL 5027,	*Typical Extinguishing and Protection Regulation of Electrical Equipment*
DL/T 5158,	*Technical Code for Meteorological Survey in Electric Power Engineering*
DL/T 5222,	*Design Technical Rule for Selecting Conductor and Electrical Equipment*
DL/T 5228,	*Code for Design of AC 110 kV ~ 500 kV Power Cable Systems for Hydro-Power Station*
JB/T 10088,	*Sound Level for 6 kV ~ 1000 kV Power Transformers*